Spon's Estimating Costs Guide to Electrical Works

Project costs at a glance

Second edition

Bryan Spain

Spon's Contractors' Handbooks

LONDON AND NEW YORK

First published 1999 by Spon Press
Second edition 2004
11 New Fetter Lane, London EC4P 4EE

Simultaneously published in the USA and Canada
by Spon Press
29 West 35th Street, New York, NY 10001

Spon Press is an imprint of the Taylor & Francis Group

Publisher's Note
This book has been prepared from camera-ready copy supplied by the author.

Printed and bound in Great Britain by TJ International Ltd, Padstow, Cornwall.

British Library Cataloguing in Publication Data
A catalogue record for this book is available from the British Library

Library of Congress Cataloging in Publication Data
A catalog record for this book has been requested

ISBN 0-415-31853-X

Spon's Estimating Costs Guide to Electrical Works

Other books by Bryan Spain

also available from Spon Press

Spon's Estimating Costs Guide to Plumbing and Heating
(2004 edition)
Pb: 0-415-31855-6

Spon's Estimating Costs Guide to Minor Works, Alterations and Repairs to Fire, Flood, Gale and Theft Damage
(2004 edition)
Pb: 0-415-31854-8

Spon's House Improvement Price Book: house extensions, storm damage work, alterations, loft conversions and insulation (2003 edition)
Pb: 0-415-30938-7

Spon's First Stage Estimating Price Book (2000 edition)
Pb: 0-415-23436-0

Spon's Construction Resource Handbook (1998 edition)
Hb: 0-419-23680-5

Information and ordering details

For price availability and ordering visit our website **www.sponpress.com**

Alternatively our books are available from all good bookshops.

Contents

Preface

The success of the previous edition of this book confirms the view that electrical contractors operating in the domestic and light industrial market need support when preparing estimates. This book provides unit rates for a wide range of items together with costed examples of new electrical work, upgrading and re-wiring to a total of 15 different sized dwellings. These project costs provide examples that can be used by electrical estimators to prepare their bids and save valuable estimating time.

I have received a great deal of support in the research necessary for this type of book. I am indebted to Mark Loughrey of Youds, Ellison & Co., Chartered Accountants of Hoylake (tel: 0151-632 3298 or www.yesl.uk.com), who are specialists in advising small construction businesses. His research for the information in the business section is based on tax legislation in force in December 2003.

I am grateful to suppliers and manufacturers in the electrical industry for their help in providing the cost information that appears in this book including Brian Carrington of Newey and Eyre. I am particularly indebted to Tony Parry who has helped me in every stage of the planning, provision of technical data and general supervision of the book's contents.

Although every care has been taken in the preparation of the book, neither the publishers nor I can accept any responsibility for the use of the information made by any individual or firm. Finally, I would welcome any constructive criticism of the book's contents and suggestions that could be incorporated into the next edition.

Bryan Spain
www.costofdiy.com
December 2003

Introduction

The contents of this book cover unit rates, project costs and general advice on business matters. The unit rates section presents analytical rates for all types of work encountered in small- to medium-sized electrical contracts. The project costs section contains the total costs for carrying out work in domestic installations in the following categories and each category is sub-divided into a different sized project:

- upgrading (testing and remedial work)
- rewiring
- new installation.

The rates used in the project costs section may vary slightly from those in the unit rates part of the book because of differing circumstances. The business section covers advice on starting and running a business together with information on taxation and VAT matters.

Materials

It should be noted that the discounts used on the materials in this book are small, particularly for cabling, where a nominal discount of 30% to 40% has been used. Much larger discounts can be obtained for large orders.

Labour

The net labour rate has been taken throughout the book at £15.50 per hour based upon the current National Standard Rates promulgated by the Joint Industry Board for the Electrical Contracting Industry (JIB) payable to Approved Electricians including a £25 weekly bonus.

Overheads and profit

This has been set at 15% for all grades of work and is deemed to cover head office and site overheads including:

- heating
- lighting
- rent
- rates
- telephones
- secretarial services
- insurances
- finance charges
- transport
- small tools
- ladders
- scaffolding etc.

Part One

UNIT RATES

Y60 Conduit and cable trunking

Y61 HV/LV cables and wiring

Y71 LV switchgear and distribution boards

Y73 Luminaires and lamps

Y74 Accessories for electrical services

Y41 Fans

Y80 Earthing and bonding

V51 Local electric heating units

W30 Data transmission

W41 Security

W50 Fire detection and alarm

Builder's work in connection with electrical installation

Sundry work

	Unit	Hours	Hours £	Mat'ls £	O & P £	Total £
Y60 CONDUIT AND CABLE TRUNKING						
Conduits						
Steel conduits surface fixed to spacer bar saddles measured elsewhere, (supplied in 3.75 metre lengths and complete with 1 coupler)						
black enamel (Class 2); heavy gauge (HGSW)						
20mm diameter	m	0.307	4.76	0.87	0.84	6.47
25mm diameter	m	0.311	4.82	1.05	0.88	6.75
32mm diameter	m	0.384	5.95	1.47	1.11	8.54
1.1/2in diameter	m	0.450	6.98	2.36	1.40	10.74
2in diameter	m	0.600	9.30	3.82	1.97	15.09
Steel or malleable iron, black enamel conduit accessories including jointing conduit to fittings						
couplings solid						
20mm diameter	each	0.008	0.12	0.20	0.05	0.37
25mm diameter	each	0.008	0.12	0.22	0.05	0.40
32mm diameter	each	0.016	0.25	0.67	0.14	1.06
1.1/2in diameter	each	0.016	0.25	0.90	0.17	1.32
2in diameter	each	0.016	0.25	1.47	0.26	1.98
reducers						
25-20mm diameter	each	0.025	0.39	2.91	0.49	3.79
32-25mm diameter	each	0.041	0.64	3.46	0.61	4.71
1.1/2in-32mm diameter	each	0.041	0.64	5.19	0.87	6.70
2in-1.1/2" diameter	each	0.041	0.64	11.37	1.80	13.81
saddles, plain, to backgrounds requiring fixings						
20mm diameter	each	0.116	1.80	0.18	0.30	2.27
25mm diameter	each	0.116	1.80	0.20	0.30	2.30

	Unit	Hours	Hours £	Mat'ls £	O & P £	Total £
Accessories (cont'd)						
spacer bar saddles; to backgrounds requiring fixings						
20mm diameter	each	0.116	1.80	0.22	0.30	2.32
25mm diameter	each	0.116	1.80	0.26	0.31	2.37
32mm diameter	each	0.116	1.80	0.69	0.37	2.86
1.1/2in diameter	each	0.116	1.80	0.71	0.38	2.88
distance spacing saddles; to backgrounds requiring fixings						
20mm diameter	each	0.125	1.94	0.73	0.40	3.07
25mm diameter	each	0.125	1.94	0.93	0.43	3.30
32mm diameter	each	0.125	1.94	1.81	0.56	4.31
Malleable iron, black enamel, small circular conduit boxes including jointing conduit to fittings; to backgrounds requiring fixings						
terminal box						
20mm diameter	each	0.100	1.55	1.54	0.46	3.55
25mm diameter	each	0.100	1.55	2.13	0.55	4.23
through box						
20mm diameter	each	0.116	1.80	1.84	0.55	4.18
25mm diameter	each	0.116	1.80	2.54	0.65	4.99
angle box						
20mm diameter	each	0.116	1.80	3.86	0.85	6.51
25mm diameter	each	0.116	1.80	4.44	0.94	7.17
tee box						
20mm diameter	each	0.150	2.33	2.02	0.65	5.00
25mm diameter	each	0.150	2.33	2.68	0.75	5.76
4-way box						
20mm diameter	each	0.183	2.84	2.55	0.81	6.19
25mm diameter	each	0.183	2.84	3.74	0.99	7.56

	Unit	Hours	Hours £	Mat'ls £	O & P £	Total £
terminal and back outlet box						
20mm diameter	each	0.150	2.33	3.98	0.95	7.25
25mm diameter	each	0.150	2.33	5.80	1.22	9.34
through and back outlet box						
20mm diameter	each	0.166	2.17	4.68	1.03	7.88
25mm diameter	each	0.166	2.17	6.41	1.29	9.87
angle and back outlet box						
20mm diameter	each	0.166	2.57	4.71	1.09	8.38
25mm diameter	each	0.166	2.57	6.41	1.35	10.33
4-way and back outlet box						
20mm diameter	each	0.233	3.61	5.21	1.32	10.14
25mm diameter	each	0.233	3.61	6.97	1.59	12.17
branch U; 2-way box						
20mm diameter	each	0.116	1.80	2.61	0.66	5.07
25mm diameter	each	0.116	1.80	5.60	1.11	8.51
branch Y; 3-way box						
20mm diameter	each	0.150	2.33	3.77	0.91	7.01
25mm diameter	each	0.150	2.33	6.25	1.29	9.86
twin through H box						
20mm diameter	each	0.183	2.84	5.33	1.22	9.39
25mm diameter	each	0.183	2.84	7.27	1.52	11.62
Conduit box covers complete with screws	each	0.025	0.39	0.35	0.11	0.85
Hot dipped galvanised (Class 4); heavy gauge conduit (HGSW)						
20mm diameter	m	0.307	4.76	1.11	0.88	6.75
25mm diameter	m	0.311	4.82	1.39	0.93	7.14
32mm diameter	m	0.384	5.95	1.94	1.18	9.08
1.1/2in diameter	m	0.450	6.98	3.04	1.50	11.52
2in diameter	m	0.600	9.30	4.59	2.08	15.97

	Unit	Hours	Hours £	Mat'ls £	O & P £	Total £
Steel or malleable iron, galvanised conduit accessories including jointing conduit to fittings						
couplings solid						
20mm diameter	each	0.008	0.12	0.23	0.05	0.41
25mm diameter	each	0.008	0.12	0.28	0.06	0.46
32mm diameter	each	0.016	0.25	0.82	0.16	1.23
1.1/2in diameter	each	0.016	0.25	1.20	0.22	1.67
2in diameter	each	0.016	0.25	1.91	0.32	2.48
reducers						
25-20mm diameter	each	0.025	0.39	1.07	0.22	1.68
32-25mm diameter	each	0.041	0.64	1.26	0.28	2.18
1.1/2in-32mm diameter	each	0.041	0.64	1.85	0.37	2.86
2-1.1/2in diameter	each	0.041	0.64	4.04	0.70	5.38
saddles, plain; to backgrounds requiring fixings						
20mm diameter	each	0.116	1.80	0.18	0.30	2.27
25mm diameter	each	0.116	1.80	0.21	0.30	2.31
32mm diameter	each	0.116	1.80	0.60	0.36	2.76
1.1/2in diameter	each	0.116	1.80	0.76	0.38	2.94
2in diameter	each	0.116	1.80	0.81	0.39	3.00
spacer bar saddles; to backgrounds requiring fixings						
20mm diameter	each	0.116	1.80	0.26	0.31	2.37
25mm diameter	each	0.116	1.80	0.30	0.31	2.41
32mm diameter	each	0.116	1.80	0.77	0.39	2.95
1.1/2in diameter	each	0.116	1.80	0.81	0.39	3.00
2in diameter	each	0.116	1.80	1.08	0.43	3.31
distance spacing saddles; to backgrounds requiring fixings						
20mm diameter	each	0.125	1.94	0.87	0.42	3.23
25mm diameter	each	0.125	1.94	1.14	0.46	3.54
32mm diameter	each	0.125	1.94	1.93	0.58	4.45
1.1/2in diameter	each	0.125	1.94	2.81	0.71	5.46
2in diameter	each	0.125	1.94	3.73	0.85	6.52

	Unit	Hours	Hours £	Mat'ls £	O & P £	Total £
Malleable iron, galvanised, small circular conduit boxes including jointing conduit to fittings; to backgrounds requiring fixings						
terminal box						
20mm diameter	each	0.100	1.55	1.65	0.48	3.68
25mm diameter	each	0.100	1.55	2.29	0.58	4.42
through box						
20mm diameter	each	0.116	1.80	1.97	0.57	4.33
25mm diameter	each	0.116	1.80	2.72	0.68	5.20
angle box						
20mm diameter	each	0.116	1.80	4.06	0.88	6.74
25mm diameter	each	0.116	1.80	5.68	1.12	8.60
tee box						
20mm diameter	each	0.150	2.33	2.15	0.67	5.15
25mm diameter	each	0.150	2.33	2.99	0.80	6.11
4-way box						
20mm diameter	each	0.183	2.84	2.65	0.82	6.31
25mm diameter	each	0.183	2.84	3.78	0.99	7.61
terminal and back outlet box						
20mm diameter	each	0.150	2.33	4.13	0.97	7.42
25mm diameter	each	0.150	2.33	5.80	1.22	9.34
through and back outlet box						
20mm diameter	each	0.166	2.57	4.75	1.10	8.42
25mm diameter	each	0.166	2.57	6.51	1.36	10.45
angle and back outlet box						
20mm diameter	each	0.166	2.57	4.75	1.10	8.42
25mm diameter	each	0.166	2.57	6.51	1.36	10.45

	Unit	Hours	Hours £	Mat'ls £	O & P £	Total £
Conduit boxes (cont'd)						
4-way and back outlet box						
20mm diameter	each	0.233	3.61	5.30	1.34	10.25
25mm diameter	each	0.233	3.61	7.08	1.60	12.30
branch U; 2-way box						
20mm diameter	each	0.116	1.80	2.79	0.69	5.28
25mm diameter	each	0.116	1.80	5.67	1.12	8.59
branch Y; 3-way box						
20mm diameter	each	0.150	2.33	4.05	0.96	7.33
25mm diameter	each	0.150	2.33	6.32	1.30	9.94
twin through H box						
20mm diameter	each	0.183	2.84	5.40	1.24	9.47
25mm diameter	each	0.183	2.84	7.27	1.52	11.62
Conduit box covers complete with screws	each	0.025	0.39	0.58	0.15	1.11
PVC conduits, surface fixed to spacer bar saddles measured elsewhere, (supplied in 3 metre lengths)						
white light gauge; super high impact						
16mm diameter	m	0.083	1.29	0.52	0.27	2.08
20mm diameter	m	0.083	1.29	0.52	0.27	2.08
25mm diameter	m	0.117	1.81	0.86	0.40	3.07
32mm diameter	m	0.117	1.81	0.96	0.42	3.19
white/black heavy gauge; super high impact						
16mm diameter	m	0.100	1.55	0.68	0.33	2.56
20mm diameter	m	0.100	1.55	0.44	0.30	2.29
25mm diameter	m	0.117	1.81	0.56	0.36	2.73
32mm diameter	m	0.117	1.81	1.68	0.52	4.02
38mm diameter	m	0.117	1.81	2.22	0.61	4.64
50mm diameter	m	0.133	2.06	3.68	0.86	6.60

	Unit	Hours	Hours £	Mat'ls £	O & P £	Total £
white/black heavy gauge; super high impact						
16mm diameter	m	0.090	1.40	0.78	0.33	2.50
20mm diameter	m	0.100	1.55	0.44	0.30	2.29
25mm diameter	m	0.117	1.81	0.56	0.36	2.73
PVC round rigid conduit accessories including cutting and jointing conduit to fittings with adhesive						
standard couplings						
16mm diameter	each	0.025	0.39	0.26	0.10	0.74
20mm diameter	each	0.025	0.39	0.26	0.10	0.74
25mm diameter	each	0.025	0.39	0.35	0.11	0.85
32mm diameter	each	0.041	0.64	0.71	0.20	1.55
38mm diameter	each	0.041	0.64	1.39	0.30	2.33
50mm diameter	each	0.041	0.64	2.44	0.46	3.54
reducers						
20-16mm diameter	each	0.025	0.39	0.53	0.14	1.06
25-20mm diameter	each	0.025	0.39	0.54	0.14	1.07
32-25mm diameter	each	0.041	0.64	0.76	0.21	1.60
standard plain bends						
20mm diameter	each	0.066	1.02	0.96	0.30	2.28
25mm diameter	each	0.066	1.02	1.25	0.34	2.61
32mm diameter	each	0.083	1.29	1.56	0.43	3.27
38mm diameter	each	0.083	1.29	3.92	0.78	5.99
50mm diameter	each	0.083	1.29	7.01	1.24	9.54
adaptors, female threaded with male bush						
20mm diameter	each	0.216	3.35	0.28	0.54	4.17
25mm diameter	each	0.250	3.88	0.46	0.65	4.99
32mm diameter	each	0.250	3.88	0.86	0.71	5.45
38mm diameter	each	0.285	4.42	1.81	0.93	7.16
50mm diameter	each	0.320	4.96	3.15	1.22	9.33

	Unit	Hours	Hours £	Mat'ls £	O & P £	Total £
Accessories (cont'd)						
adaptors, male threaded with lockrings						
20mm diameter	each	0.216	3.35	0.31	0.55	4.21
25mm diameter	each	0.250	3.88	0.49	0.65	5.02
32mm diameter	each	0.250	3.88	0.89	0.71	5.48
male bushes screwed						
20mm diameter	each	0.108	1.67	0.12	0.27	2.06
25mm diameter	each	0.125	1.94	0.17	0.32	2.42
32mm diameter	each	0.125	1.94	0.26	0.33	2.53
38mm diameter	each	0.142	2.20	0.44	0.40	3.04
50mm diameter	each	0.160	2.48	0.94	0.51	3.93
bellmouth bushes (white only)						
20mm diameter	each	0.108	1.67	0.15	0.27	2.10
25mm diameter	each	0.125	1.94	0.21	0.32	2.47
sleeves for bellmouth bushes (white only)						
20mm diameter	each	0.108	1.67	0.19	0.28	2.14
25mm diameter	each	0.125	1.94	0.34	0.34	2.62
lockrings screwed						
20mm diameter	each	0.108	1.67	0.25	0.29	2.21
25mm diameter	each	0.125	1.94	0.28	0.33	2.55
32mm diameter	each	0.125	1.94	0.33	0.34	2.61
strap saddles; to backgrounds requiring fixings						
20mm diameter	each	0.116	1.80	0.15	0.29	2.24
25mm diameter	each	0.116	1.80	0.19	0.30	2.29
32mm diameter	each	0.116	1.80	0.25	0.31	2.36

	Unit	Hours	Hours £	Mat'ls £	O & P £	Total £
spacer bar saddles; to backgrounds requiring fixings						
16mm diameter	each	0.116	1.80	0.33	0.32	2.45
20mm diameter	each	0.116	1.80	0.43	0.33	2.56
25mm diameter	each	0.116	1.80	0.43	0.33	2.56
32mm diameter	each	0.116	1.80	0.87	0.40	3.07
38mm diameter	each	0.116	1.80	1.56	0.50	3.86
50mm diameter	each	0.116	1.80	2.64	0.67	5.10
round spring saddle clips, centre fixing; to backgrounds requiring fixings						
20mm diameter	each	0.143	2.22	0.29	0.38	2.88
25mm diameter	each	0.175	2.71	0.47	0.48	3.66
PVC circular junction boxes including cutting and jointing conduits to fittings with adhesive; to backgrounds requiring fixings						
back outlet box						
20mm diameter	each	0.050	0.78	1.35	0.32	2.44
25mm diameter	each	0.050	0.78	1.57	0.35	2.70
terminal box						
20mm diameter	each	0.050	0.78	1.00	0.27	2.04
25mm diameter	each	0.050	0.78	1.57	0.35	2.70
through box						
20mm diameter	each	0.066	1.02	1.11	0.32	2.45
25mm diameter	each	0.066	1.02	1.71	0.41	3.14
angle box						
20mm diameter	each	0.066	1.02	1.11	0.32	2.45
25mm diameter	each	0.066	1.02	1.71	0.41	3.14
3-way tee box						
20mm diameter	each	0.116	1.80	1.20	0.45	3.45
25mm diameter	each	0.116	1.80	1.88	0.55	4.23

	Unit	Hours	Hours £	Mat'ls £	O & P £	Total £
Junction boxes (cont'd)						
4-way intersection box						
20mm diameter	each	0.183	2.84	1.42	0.64	4.89
25mm diameter	each	0.183	2.84	2.13	0.74	5.71
terminal and back outlet box						
20mm diameter	each	0.066	1.02	1.56	0.39	2.97
25mm diameter	each	0.066	1.02	2.04	0.46	3.52
through and back outlet box						
20mm diameter	each	0.082	1.27	1.68	0.44	3.39
25mm diameter	each	0.082	1.27	2.32	0.54	4.13
angle and back outlet box						
20mm diameter	each	0.082	1.27	1.68	0.44	3.39
25mm diameter	each	0.082	1.27	2.32	0.54	4.13
3-way and back outlet box						
20mm diameter	each	0.132	2.05	1.88	0.59	4.51
25mm diameter	each	0.132	2.05	2.48	0.68	5.20
4-way intersection and back outlet box						
20mm diameter	each	0.199	3.08	1.91	0.75	5.74
25mm diameter	each	0.199	3.08	2.68	0.86	6.63
PVC circular box accessories; to backgrounds requiring fixings						
extension rings						
12.5mm deep	each	0.066	1.02	0.67	0.25	1.95
20mm deep	each	0.066	1.02	0.87	0.28	2.18
25mm deep	each	0.066	1.02	1.04	0.31	2.37
38mm deep	each	0.066	1.02	1.10	0.32	2.44
50mm deep	each	0.066	1.02	1.44	1.22	3.68

	Unit	Hours	Hours £	Mat'ls £	O & P £	Total £
circular lids and gaskets						
standard 65mm diameter	each	0.025	0.39	0.28	1.22	1.89
overlapping 85mm diameter	each	0.025	0.39	0.51	1.22	2.12
gasket 65mm diameter	each	0.025	0.39	0.34	1.22	1.95
PVC oval conduit, surface fixed to oval spring saddle clips measured elsewhere, (supplied in 3 metre lengths)						
white super high impact						
12mm diameter	m	0.020	0.31	0.36	0.10	0.77
16mm diameter	m	0.020	0.31	0.44	0.11	0.86
20mm diameter	m	0.070	1.09	0.57	0.25	1.90
25mm diameter	m	0.070	1.09	0.79	0.28	2.16
32mm diameter	m	0.200	3.10	1.37	0.67	5.14
38mm diameter	m	0.200	3.10	2.74	0.88	6.72
PVC oval conduit accessories; including cutting and jointing conduit to fittings with adhesive						
oval to round adaptors						
20mm diameter	each	0.108	1.67	0.91	1.22	3.80
oval spring saddle clips; to backgrounds requiring fixings						
12mm diameter	each	0.080	1.24	0.13	0.21	1.58
16mm diameter	each	0.080	1.24	0.13	0.21	1.58
20mm diameter	each	0.110	1.71	0.14	0.28	2.12
25mm diameter	each	0.145	2.25	0.14	0.36	2.75

	Unit	Hours	Hours £	Mat'ls £	O & P £	Total £
PVC channelling flush mounted; to backgrounds requiring fixings (supplied in 2 metre lengths)						
white light gauge; super high impact						
12mm diameter	each	0.066	1.02	0.33	0.20	1.56
25mm diameter	each	0.066	1.02	0.48	0.23	1.73
38mm diameter	each	0.100	1.55	0.71	0.34	2.60
white standard gauge; super high impact						
12mm diameter	each	0.066	1.02	0.38	0.21	1.61
25mm diameter	each	0.066	1.02	0.53	0.23	1.79
38mm diameter	each	0.100	1.55	0.75	0.35	2.65

Trunking

	Unit	Hours	Hours £	Mat'ls £	O & P £	Total £
Galvanised steel trunking with return edge including cover and internal sleeve coupling and earth continuity strap; to backgrounds requiring fixings (supplied in 3 metre lengths)						
single compartment						
50 × 50mm	m	0.283	4.39	2.67	1.06	8.11
75 × 75mm	m	0.350	5.43	3.53	1.34	10.30
100 × 100mm	m	0.466	7.22	4.75	1.80	13.77
150 × 150mm	m	0.616	9.55	7.00	2.48	19.03
two compartment						
50 × 50mm	m	0.316	4.90	6.85	1.76	13.51
75 × 75mm	m	0.383	5.94	9.65	2.34	17.92
100 × 100mm	m	0.499	7.73	12.46	3.03	23.22
150 × 150mm	m	0.649	10.06	21.94	4.80	36.80

	Unit	Hours	Hours £	Mat'ls £	O & P £	Total £
three compartment						
150 × 50mm	m	0.616	9.55	11.81	3.20	24.56
150 × 75mm	m	0.616	9.55	12.77	3.35	25.67
150 × 100mm	m	0.633	9.81	14.67	3.67	28.15
150 × 150mm	m	0.666	10.32	22.51	4.92	37.76
Galvanised steel trunking single compartment fitttings including cutting and jointing trunking to fittings						
additional couplers						
50 × 50mm	each	0.150	2.33	1.21	0.53	4.07
75 × 75mm	each	0.150	2.33	1.42	0.56	4.31
100 × 100mm	each	0.183	2.84	1.73	0.68	5.25
150 × 150mm	each	0.233	3.61	2.28	0.88	6.78
equal tees						
50 × 50mm	each	0.216	3.35	4.04	1.11	8.50
75 × 75mm	each	0.216	3.35	4.73	1.21	9.29
100 × 100mm	each	0.250	3.88	5.78	1.45	11.10
150 × 150mm	each	0.316	4.90	9.56	2.17	16.63
90 degrees elbows (bends)						
50 × 50mm	each	0.133	2.06	4.04	0.92	7.02
75 × 75mm	each	0.133	2.06	4.73	1.02	7.81
100 × 100mm	each	0.166	2.57	5.78	1.25	9.61
150 × 150mm	each	0.233	3.61	9.56	1.98	15.15
45 degrees elbows (bends)						
50 × 50mm	each	0.133	2.06	4.04	0.92	7.02
75 × 75mm	each	0.133	2.06	4.73	1.02	7.81
100 × 100mm	each	0.166	2.57	5.78	1.25	9.61
150 × 150mm	each	0.233	3.61	9.56	1.98	15.15
four way (cross overs)						
50 × 50mm	each	0.283	4.39	6.56	1.64	12.59
75 × 75mm	each	0.283	4.39	8.35	1.91	14.65
100 × 100mm	each	0.350	5.43	9.77	2.28	17.47
150 × 150mm	each	0.416	6.45	12.86	2.90	22.20

	Unit	Hours	Hours £	Mat'ls £	O & P £	Total £
Fittings (cont'd)						
reducers						
75 × 75 - 50 × 50mm	each	0.166	2.57	7.30	1.48	11.35
100 × 100 - 50 × 50mm	each	0.183	2.84	8.66	1.72	13.22
100 × 100 - 75 × 75mm	each	0.183	2.84	8.66	1.72	13.22
150 × 150 - 50 × 50mm	each	0.200	3.10	11.71	2.22	17.03
150 × 150 - 75 × 75mm	each	0.200	3.10	11.71	2.22	17.03
150 × 150 - 100 × 100mm	each	0.200	3.10	11.71	2.22	17.03
stop ends						
50 × 50mm	each	0.050	0.78	1.00	0.27	2.04
75 × 75mm	each	0.050	0.78	1.37	0.32	2.47
100 × 100mm	each	0.066	1.02	1.63	0.40	3.05
150 × 150mm	each	0.100	1.55	2.36	0.59	4.50
cable retainers						
50 × 50mm	each	0.016	0.25	0.63	0.13	1.01
75 × 75mm	each	0.016	0.25	0.68	0.14	1.07
100 × 100mm	each	0.016	0.25	0.74	0.15	1.14
150 × 150mm	each	0.016	0.25	0.89	0.17	1.31
flange couplers						
50 × 50mm	each	0.200	3.10	1.37	0.67	5.14
75 × 75mm	each	0.200	3.10	1.73	0.72	5.55
100 × 100mm	each	0.250	3.88	2.00	0.88	6.76
150 × 150mm	each	0.333	5.16	2.99	1.22	9.37
Galvanised steel trunking two compartment fitttings including cutting and jointing trunking to fittings						
equal tees						
50 × 50mm	each	0.249	3.86	6.85	1.61	12.32
75 × 75mm	each	0.249	3.86	8.13	1.80	13.79
100 × 100mm	each	0.283	4.39	9.27	2.05	15.70
150 × 150mm	each	0.349	5.41	14.67	3.01	23.09

	Unit	Hours	Hours £	Mat'ls £	O & P £	Total £
90 degrees elbows (bends)						
50 × 50mm	each	0.166	2.57	5.20	1.17	8.94
75 × 75mm	each	0.166	2.57	6.38	1.34	10.30
100 × 100mm	each	0.199	3.08	8.08	1.67	12.84
150 × 150mm	each	0.266	4.12	11.94	2.41	18.47
45 degrees elbows (bends)						
50 × 50mm	each	0.166	2.57	5.53	1.22	9.32
75 × 75mm	each	0.166	2.57	7.13	1.46	11.16
100 × 100mm	each	0.199	3.08	9.03	1.82	13.93
150 × 150mm	each	0.266	4.12	13.35	2.62	20.09
four way (cross overs)						
50 × 50mm	each	0.316	4.90	9.37	2.14	16.41
75 × 75mm	each	0.316	4.90	12.61	2.63	20.13
100 × 100mm	each	0.383	5.94	14.72	3.10	23.75
150 × 150mm	each	0.449	6.96	20.69	4.15	31.80
reducers						
75 × 75 - 50 × 50mm	each	0.199	3.08	7.46	1.58	12.13
100 × 100 - 50 × 50mm	each	0.216	3.35	8.44	1.77	13.56
100 × 100 - 75 × 75mm	each	0.216	3.35	8.44	1.77	13.56
150 × 150 - 50 × 50mm	each	0.233	3.61	18.87	3.37	25.85
150 × 150 - 75 × 75mm	each	0.233	3.61	18.87	3.37	25.85
150 × 150 - 100 × 100mm	each	0.233	3.61	18.87	3.37	25.85
Galvanised steel trunking three compartment fitttings including cutting and jointing trunking to fittings						
equal tee						
150 × 50mm	each	0.333	5.16	11.38	2.48	19.02
150 × 75mm	each	0.333	5.16	13.34	2.78	21.28
150 × 100mm	each	0.333	5.16	19.55	3.71	28.42
150 × 150mm	each	0.346	5.36	30.01	5.31	40.68

	Unit	Hours	Hours £	Mat'ls £	O & P £	Total £
90 degree elbows (bends)						
150 × 50mm	each	0.283	4.39	8.81	1.98	15.18
150 × 75mm	each	0.283	4.39	12.77	2.57	19.73
150 × 100mm	each	0.283	4.39	14.67	2.86	21.91
150 × 150mm	each	0.283	4.39	22.51	4.03	30.93
45 degree elbows (bends)						
150 × 50mm	each	0.283	4.39	9.37	2.06	15.82
150 × 75mm	each	0.283	4.39	12.77	2.57	19.73
150 × 100mm	each	0.283	4.39	14.67	2.86	21.91
150 × 150mm	each	0.283	4.39	22.51	4.03	30.93
four way (cross overs)						
150 × 50mm	each	0.450	6.98	17.20	3.63	27.80
150 × 75mm	each	0.450	6.98	19.06	3.91	29.94
150 × 100mm	each	0.450	6.98	21.89	4.33	33.19
150 × 150mm	each	0.466	7.22	32.62	5.98	45.82
reducers						
150 × 50 - 50 × 50mm	each	0.250	3.88	8.80	1.90	14.58
150 × 75 - 50 × 50mm	each	0.250	3.88	15.96	2.98	22.81
150 × 100 - 100 × 100mm	each	0.250	3.88	18.33	3.33	25.54
150 × 150 - 50 × 50mm	each	0.250	3.88	27.32	4.68	35.87
150 × 150 - 75 × 75mm	each	0.250	3.88	27.32	4.68	35.87
150 × 150 - 100 × 100mm	each	0.250	3.88	27.32	4.68	35.87
Copper earth link including bolts and washers and fixing to trunking and drilling						
copper earth link	each	0.083	1.29	0.33	0.24	1.86

Lighting trunking

Hot dipped pre-coated galvanised steel lighting trunking body; to backgrounds requiring fixings (supplied in 3.5 metre lengths)

	Unit	Hours	Hours £	Mat'ls £	O & P £	Total £
50 × 50mm	m	0.283	4.39	1.88	0.94	7.21

	Unit	Hours	Hours £	Mat'ls £	O & P £	Total £
Galvanised steel lighting trunking accessories including cutting and jointing trunking to fittings						
couplings (330mm long)						
50 × 50mm	m	0.150	2.33	1.31	0.55	4.18
equal tees						
50 × 50mm	m	0.216	3.35	3.73	1.06	8.14
90 degree elbows (bends)						
50 × 50mm	m	0.133	2.06	3.52	0.84	6.42
four way (cross overs)						
50 × 50mm	m	0.283	4.39	6.46	1.63	12.47
end caps						
50 × 50mm	m	0.050	0.78	0.47	0.19	1.43
suspension brackets						
50 × 50mm	m	0.166	2.57	0.74	0.50	3.81
cable retainers						
50 × 50mm	m	0.016	0.25	0.47	0.11	0.83
attachment assembly						
50 × 50mm; comprising latch and clamp plates with smooth bore 20mm diameter brass bush and locknut	m	0.333	5.16	1.10	0.94	7.20

	Unit	Hours	Hours £	Mat'ls £	O & P £	Total £
Lighting trunking cover (supplied in 1.75 metre lengths)						
50 × 50mm galvanised	m	0.050	0.78	0.96	0.26	2.00
50 × 50mm PVC	m	0.050	0.78	1.74	0.38	2.89
PVC heavy duty cable trunking complete with cover plate; to backgrounds requiring fixings (supplied in 3 metre lengths)						
single compartment						
50 × 50mm	m	0.150	2.33	7.41	1.46	11.20
50 × 75mm	m	0.183	2.84	8.97	1.77	13.58
50 × 100mm	m	0.208	3.22	12.17	2.31	17.70
50 × 150mm	m	0.208	3.22	19.28	3.38	25.88
75 × 75mm	m	0.183	2.84	11.10	2.09	16.03
100 × 100mm	m	0.250	3.88	17.61	3.22	24.71
150 × 150mm	m	0.300	4.65	37.81	6.37	48.83
PVC heavy duty cable trunking single compartment fittings including cutting and jointing trunking to fittings						
internal/external tees						
50 × 50mm	each	0.100	1.55	16.64	2.73	20.92
50 × 75mm	each	0.100	1.55	17.45	2.85	21.85
50 × 100mm	each	0.100	1.55	24.15	3.86	29.56
50 × 150mm	each	0.100	1.55	45.15	7.01	53.71
75 × 75mm	each	0.100	1.55	20.22	3.27	25.04
100 × 100mm	each	0.100	1.55	28.88	4.56	34.99
150 × 150mm	each	0.166	2.57	76.33	11.84	90.74

	Unit	Hours	Hours £	Mat'ls £	O & P £	Total £
flat angles (90 degree bends)						
50 × 100mm	each	0.066	1.02	15.73	2.51	19.27
50 × 150mm	each	0.116	1.80	36.72	5.78	44.30
75 × 75mm	each	0.066	1.02	11.94	1.94	14.91
100 × 100mm	each	0.066	1.02	27.76	4.32	33.10
150 × 150mm	each	0.116	1.80	58.37	9.03	69.19
reducing couplings						
75 × 75 - 50 × 50mm	each	0.100	1.55	8.60	1.52	11.67
100 × 100 - 75 × 75mm	each	0.100	1.55	10.14	1.75	13.44
150 × 150 - 100 × 100mm	each	0.133	2.06	11.93	2.10	16.09
flanged couplings						
50 × 50mm	each	0.200	3.10	1.53	0.69	5.32
50 × 75mm	each	0.200	3.10	2.08	0.78	5.96
50 × 100mm	each	0.250	3.88	4.22	1.21	9.31
50 × 150mm	each	0.333	5.16	9.60	2.21	16.98
75 × 75mm	each	0.200	3.10	3.61	1.01	7.72
100 × 100mm	each	0.250	3.88	6.87	1.61	12.36
stop ends						
50 × 50mm	each	0.033	0.51	1.04	0.23	1.78
50 × 75mm	each	0.033	0.51	1.56	0.31	2.38
50 × 100mm	each	0.050	0.78	2.38	0.47	3.63
50 × 150mm	each	0.083	1.29	9.99	1.69	12.97
75 × 75mm	each	0.033	0.51	2.08	0.39	2.98
100 × 100mm	each	0.050	0.78	4.43	0.78	5.99
150 × 150mm	each	0.083	1.29	14.73	2.40	18.42
UPVC extra super high impact grade 'Mini' trunking; to backgrounds requiring fixings (supplied in 3 metre lengths)						
single compartment						
16 × 10mm	m	0.116	1.80	1.24	0.46	3.49
16 × 16mm	m	0.133	2.06	2.05	0.62	4.73
16 × 25mm	m	0.133	2.06	2.31	0.66	5.03
16 × 38mm	m	0.133	2.06	2.80	0.73	5.59

	Unit	Hours	Hours £	Mat'ls £	O & P £	Total £
Fittings (cont'd)						
25 × 38mm	m	0.150	2.33	3.38	0.86	6.56
50 × 25mm	m	0.150	2.33	4.25	0.99	7.56
38 × 38mm	m	0.150	2.33	5.10	1.11	8.54
UPVC extra super high impact grade single compartment fittings including cutting and jointing trunking to fittings						
equal tees						
16 × 25mm	each	0.024	0.37	1.13	0.23	1.73
16 × 38mm	each	0.024	0.37	1.13	0.23	1.73
25 × 38mm	each	0.027	0.42	1.62	0.31	2.34
50 × 25mm	each	0.027	0.42	2.46	0.43	3.31
38 × 38mm	each	0.027	0.42	2.28	0.40	3.10
bends 90 degrees						
16 × 25mm	each	0.016	0.25	0.67	0.14	1.06
16 × 38mm	each	0.016	0.25	0.67	0.14	1.06
25 × 38mm	each	0.018	0.28	1.63	0.29	2.20
50 × 25mm	each	0.018	0.28	2.45	0.41	3.14
38 × 38mm	each	0.018	0.28	2.33	0.39	3.00
couplers						
16 × 25mm	each	0.016	0.25	0.67	0.14	1.06
16 × 38mm	each	0.016	0.25	0.67	0.14	1.06
25 × 38mm	each	0.018	0.28	1.65	0.29	2.22
50 × 25mm	each	0.018	0.28	2.48	0.41	3.17
38 × 38mm	each	0.018	0.28	2.32	0.39	2.99
end caps						
16 × 25mm	each	0.008	0.12	0.67	0.12	0.91
16 × 38mm	each	0.008	0.12	0.67	0.12	0.91
25 × 38mm	each	0.009	0.14	0.82	0.14	1.10
50 × 25mm	each	0.009	0.14	1.20	0.20	1.54
38 × 38mm	each	0.009	0.14	1.08	0.18	1.40

	Unit	Hours	Hours £	Mat'ls £	O & P £	Total £
UPVC moulded three compartment skirting/dado trunking with chamfered top section; to backgrounds requiring fixings						
180 × 52mm						
trunking assembly comprising lid	m	0.250	3.88	24.48	4.25	32.61
external angle	each	0.150	2.33	5.64	1.19	9.16
internal angle	each	0.150	2.33	5.64	1.19	9.16
joint cover	each	0.050	0.78	3.30	0.61	4.69
fi×ing end caps	pair	0.100	1.55	3.57	0.77	5.89
flat angle unit	each	0.200	3.10	50.53	8.04	61.67
flat tee unit	each	0.225	3.49	66.04	10.43	79.96
1 gang socket box	each	0.250	3.88	2.28	0.92	7.08
2 gang socket box	each	0.250	3.88	2.76	1.00	7.63

Tray and basket

	Unit	Hours	Hours £	Mat'ls £	O & P £	Total £
Galvanised slotted pattern cable tray; to backgrounds requiring fixings; bracket support measured elsewhere (supplied in 3 metre lengths)						
light duty						
50mm wide	m	0.216	3.35	1.38	0.71	5.44
75mm wide	m	0.216	3.35	1.58	0.74	5.67
100mm wide	m	0.316	4.90	2.01	1.04	7.94
150mm wide	m	0.316	4.90	2.57	1.12	8.59
225mm wide	m	0.383	5.94	4.99	1.64	12.57
300mm wide	m	0.500	7.75	6.91	2.20	16.86
Galvanised slotted cable tray fittings including cutting and jointing tray to fittings						
flat bends						
50mm wide	m	0.200	3.10	2.52	0.84	6.46
75mm wide	m	0.200	3.10	2.73	0.87	6.70

	Unit	Hours	Hours £	Mat'ls £	O & P £	Total £
Fittings (cont'd)						
100mm wide	m	0.216	3.35	2.84	0.93	7.12
150mm wide	m	0.216	3.35	3.62	1.05	8.01
225mm wide	m	0.283	4.39	5.04	1.41	10.84
300mm wide	m	0.316	4.90	7.82	1.91	14.63
equal tees						
50mm wide	m	0.300	4.65	3.88	1.28	9.81
75mm wide	m	0.300	4.65	3.94	1.29	9.88
100mm wide	m	0.316	4.90	4.36	1.39	10.65
150mm wide	m	0.316	4.90	5.76	1.60	12.26
225mm wide	m	0.400	6.20	7.88	2.11	16.19
300mm wide	m	0.500	7.75	11.81	2.93	22.49
medium duty						
75mm wide	m	0.216	3.35	2.33	0.85	6.53
100mm wide	m	0.316	4.90	2.57	1.12	8.59
150mm wide	m	0.316	4.90	3.33	1.23	9.46
225mm wide	m	0.383	5.94	4.25	1.53	11.71
300mm wide	m	0.500	7.75	7.09	2.23	17.07
Galvanised slotted cable tray fittings including cutting and jointing tray to fittings						
adjustable flat bends 30-90 degrees						
75mm wide	m	0.200	3.10	11.66	2.21	16.97
100mm wide	m	0.216	3.35	12.76	2.42	18.52
150mm wide	m	0.216	3.35	13.81	2.57	19.73
225mm wide	m	0.283	4.39	16.54	3.14	24.07
300mm wide	m	0.316	4.90	20.74	3.85	29.48
equal tees						
75mm wide	m	0.300	4.65	16.43	3.16	24.24
100mm wide	m	0.316	4.90	16.96	3.28	25.14
150mm wide	m	0.316	4.90	19.11	3.60	27.61
225mm wide	m	0.400	6.20	20.95	4.07	31.22
300mm wide	m	0.500	7.75	27.14	5.23	40.12

	Unit	Hours	Hours £	Mat'ls £	O & P £	Total £
couplers required for medium duty straight tray only excluding bolts and nuts	pair	0.150	2.33	2.00	0.65	4.97
Electroplated zinc finish wire cable tray 'basket'; to backgrounds requiring fixings, fixings measured elsewhere (supplied in 3 metre lengths)						
30mm upstand						
30 × 50mm wide	m	0.067	1.04	2.17	0.48	3.69
30 × 100mm wide	m	0.117	1.81	2.37	0.63	4.81
30 × 150mm wide	m	0.167	2.59	2.77	0.80	6.16
30 × 200mm wide	m	0.217	3.36	3.30	1.00	7.66
30 × 300mm wide	m	0.267	4.14	4.84	1.35	10.33
54mm upstand						
54 × 50mm wide	m	0.067	1.04	2.30	0.50	3.84
54 × 100mm wide	m	0.117	1.81	3.09	0.74	5.64
54 × 150mm wide	m	0.167	2.59	3.61	0.93	7.13
54 × 200mm wide	m	0.217	3.36	4.10	1.12	8.58
54 × 300mm wide	m	0.267	4.14	5.51	1.45	11.10
Electroplated zinc finish wire cable tray 'basket' tray fittings including cutting and jointing tray to fittings						
couplings						
spring clips	each	0.008	0.12	1.07	0.18	1.37
clamp assembly comprising bolt, nut, upper and lower clamps	each	0.013	0.20	0.33	0.08	0.61
support and fixing plates						
support channel (150mm)	each	0.133	2.06	1.96	0.60	4.62
support channel (300mm)	each	0.133	2.06	3.71	0.87	6.64
channel fixing plate	each	0.133	2.06	0.54	0.39	3.00
stand off bracket	each	0.133	2.06	2.69	0.71	5.46

	Unit	Hours	Hours £	Mat'ls £	O & P £	Total £
Sundries						
Flexible conduit take-off plate	each	0.100	1.55	2.33	0.58	4.46
Tray cover, medium duty (supplied in 3 metre lengths)						
50mm wide	each	0.500	7.75	6.34	2.11	16.20
100mm wide	each	0.500	7.75	8.74	2.47	18.96
150mm wide	each	0.580	8.99	9.49	2.77	21.25
200mm wide	each	0.280	4.34	11.03	2.31	17.68
300mm wide	each	0.660	10.23	17.32	4.13	31.68
Sundry supports; to backgrounds requiring fixings						
cantilever cable tray supports; hot dipped galvanised						
50/75mm wide	each	0.167	2.59	1.58	0.63	4.79
75/100mm wide	each	0.167	2.59	1.61	0.63	4.83
100/150mm wide	each	0.167	2.59	1.80	0.66	5.05
150/225mm wide	each	0.250	3.88	1.84	0.86	6.57
225mm wide	each	0.250	3.88	3.15	1.05	8.08
300mm wide	each	0.333	5.16	3.79	1.34	10.29
vertical cable tray supports; hot dipped galvanised (Swift)						
50mm deep	each	0.083	1.29	2.55	0.58	4.41
75mm deep	each	0.083	1.29	2.62	0.59	4.49
100mm deep	each	0.083	1.29	2.77	0.61	4.66
150mm deep	each	0.083	1.29	2.85	0.62	4.76
225mm deep	each	0.083	1.29	5.47	1.01	7.77
300mm deep	each	0.083	1.29	7.36	1.30	9.94

	Unit	Hours	Hours £	Mat'ls £	O & P £	Total £
Pre-galvanised channel support system; to backgrounds requiring fixings						
channel (supplied in 3 and 6 metre lengths)						
shallow solid	m	0.175	2.71	2.02	0.71	5.44
shallow slotted	m	0.125	1.94	2.27	0.63	4.84
deep solid	m	0.183	2.84	2.51	0.80	6.15
deep slotted	m	0.133	2.06	2.98	0.76	5.80
Pre-galvanised channel accessories						
support arms; cantilever						
150mm arm	each	0.083	1.29	3.05	0.65	4.99
300mm arm	each	0.083	1.29	4.27	0.83	6.39
450mm arm	each	0.116	1.80	4.80	0.99	7.59
600mm arm	each	0.116	1.80	6.13	1.19	9.12
90 degrees brackets						
48 × 48mm	each	0.066	1.02	0.46	0.22	1.71
53 × 42mm	each	0.099	1.53	0.46	0.30	2.29
98 × 42mm	each	0.099	1.53	0.74	0.34	2.62
98 × 46mm	each	0.132	2.05	1.12	0.47	3.64
Fixings						
beam clamps Z850	each	0.050	0.78	1.16	0.29	2.23
beam clamps Z851	each	0.050	0.78	2.59	0.50	3.87
window beam clamps Z852/41	each	0.050	0.78	1.68	0.37	2.82
beam clamps Z853	each	0.100	1.55	1.51	0.46	3.52
beam clamps Z854	each	0.050	0.78	1.51	0.34	2.63
beam clamps Z855	each	0.100	1.55	1.75	0.50	3.80

	Unit	Hours	Hours £	Mat'ls £	O & P £	Total £
toe beam clamps Z856	each	0.033	0.51	0.67	0.18	1.36
shallow channel end caps	each	0.050	0.78	0.22	0.15	1.14
deep channel end caps	each	0.050	0.78	0.22	0.15	1.14
Cover plates (supplied in 3 metre lengths)						
plastic	m	0.016	0.25	0.98	0.18	1.41
Spring nuts						
M6; short	each	0.025	0.39	0.33	0.11	0.83
M6; long	each	0.033	0.51	0.33	0.13	0.97
M10; short	each	0.025	0.39	0.45	0.13	0.96
M10; long	each	0.033	0.51	0.45	0.14	1.11
Threaded rod, cutting to lengths not exceeding 300mm long (supplied in 3 metre lengths)						
6mm diameter	m	0.066	1.02	1.05	0.31	2.38
10mm diameter	m	0.083	1.29	1.54	0.42	3.25

Y61 HV/LV CABLES AND WIRING

Cables and wiring

600/1000 volt grade XLPE insulated SWA, LSF sheathed; stranded copper conductors; BS 54567 to backgrounds requiring fixings, i.e. cleats measured elsewhere

	Unit	Hours	Hours £	Mat'ls £	O & P £	Total £
2 core cable (6942XL)						
1.5mm2	m	0.083	1.29	1.21	0.37	2.87
2.5mm2	m	0.083	1.29	1.38	0.40	3.07
4.0mm2	m	0.133	2.06	2.00	0.61	4.67
6.0mm2	m	0.133	2.06	2.30	0.65	5.02
10.0mm2	m	0.175	2.71	3.10	0.87	6.68
16.0mm2	m	0.175	2.71	4.30	1.05	8.06
25.0mm2	m	0.258	4.00	2.22	0.93	7.15
35.0mm2	m	0.258	4.00	2.68	1.00	7.68
3 core cable (6943XL)						
1.5mm2	m	0.091	1.41	1.33	0.41	3.15
2.5mm2	m	0.091	1.41	1.46	0.43	3.30
4.0mm2	m	0.150	2.33	1.98	0.65	4.95
6.0mm2	m	0.150	2.33	2.71	0.76	5.79
10.0mm2	m	0.200	3.10	3.76	1.03	7.89
16.0mm2	m	0.200	3.10	5.02	1.22	9.34
25.0mm2	m	0.283	4.39	2.68	1.06	8.13
35.0mm2	m	0.283	4.39	3.43	1.17	8.99
4 core cable (6944XL)						
1.5mm2	m	0.100	1.55	1.47	0.45	3.47
2.5mm2	m	0.100	1.55	1.76	0.50	3.81
4.0mm2	m	0.166	2.57	2.63	0.78	5.98
6.0mm2	m	0.166	2.57	3.34	0.89	6.80
10.0mm2	m	0.216	3.35	4.37	1.16	8.88
16.0mm2	m	0.216	3.35	6.13	1.42	10.90
25.0mm2	m	0.316	4.90	3.24	1.22	9.36
35.0mm2	m	0.316	4.90	3.96	1.33	10.19

	Unit	Hours	Hours £	Mat'ls £	O & P £	Total £
Complete indoor type termination (BWC) cable lugs, brass gland, locknut, earthing ring, PVC cable shroud including drilling and cutting steel gland plate and stripping 300mm long tails *NB Add 0.05 hours per metre for every extra 500mm of cable tails required*						
2 core XLPE/SWA/LSF cable						
1.5mm2	each	0.483	7.49	4.69	1.83	14.00
2.5mm2	each	0.483	7.49	4.69	1.83	14.00
4.0mm2	each	0.483	7.49	4.82	1.85	14.15
6.0mm2	each	0.500	7.75	6.30	2.11	16.16
10.0mm2	each	0.650	10.08	6.42	2.47	18.97
16.0mm2	each	0.766	11.87	9.86	3.26	24.99
25.0mm2	each	0.983	15.24	9.93	3.77	28.94
35.0mm2	each	1.050	16.28	14.28	4.58	35.14
3 core XLPE/SWA/LSF cable						
1.5mm2	each	0.516	8.00	4.78	1.92	14.69
2.5mm2	each	0.516	8.00	4.78	1.92	14.69
4.0mm2	each	0.516	8.00	4.98	1.95	14.92
6.0mm2	each	0.533	8.26	6.46	2.21	16.93
10.0mm2	each	0.716	11.10	6.63	2.66	20.39
16.0mm2	each	0.833	12.91	10.09	3.45	26.45
25.0mm2	each	1.050	16.28	14.22	4.57	35.07
35.0mm2	each	1.150	17.83	13.57	4.71	36.10
4 core XLPE/SWA/LSF cable						
1.5mm2	each	0.566	8.77	4.87	2.05	15.69
2.5mm2	each	0.566	8.77	4.87	2.05	15.69
4.0mm2	each	0.566	8.77	6.62	2.31	17.70
6.0mm2	each	0.583	9.04	6.62	2.35	18.00
10.0mm2	each	0.816	12.65	10.27	3.44	26.36
16.0mm2	each	0.933	14.46	10.86	3.80	29.12

	Unit	Hours	Hours £	Mat'ls £	O & P £	Total £
4 core XLPE/SWA/LSF cable (cont'd)						
25.0mm2	each	1.150	17.83	14.49	4.85	37.16
35.0mm2	each	1.300	20.15	15.11	5.29	40.55
600/1000 volt grade XLPE insulated SWA, LSF sheathed; stranded copper conductors; BS 5467 laid in trenches including marker tape (cable warning tiles are measured elsewhere)						
2 core cable 6942XL						
1.5mm2	m	0.116	1.80	1.29	0.46	3.55
2.5mm2	m	0.116	1.80	1.46	0.49	3.75
4.0mm2	m	0.116	1.80	2.08	0.58	4.46
6.0mm2	m	0.116	1.80	2.38	0.63	4.80
10.0mm2	m	0.208	3.22	3.18	0.96	7.36
16.0mm2	m	0.208	3.22	4.38	1.14	8.74
25.0mm2	m	0.291	4.51	2.30	1.02	7.83
35.0mm2	m	0.291	4.51	2.76	1.09	8.36
3 core cable 6943XL						
1.5mm2	m	0.124	1.92	1.41	0.50	3.83
2.5mm2	m	0.124	1.92	1.54	0.52	3.98
4.0mm2	m	0.183	2.84	2.06	0.73	5.63
6.0mm2	m	0.183	2.84	2.79	0.84	6.47
10.0mm2	m	0.233	3.61	3.84	1.12	8.57
16.0mm2	m	0.233	3.61	5.10	1.31	10.02
25.0mm2	m	0.316	4.90	2.76	1.15	8.81
35.0mm2	m	0.316	4.90	3.51	1.26	9.67
4 core cable 6944XL						
1.5mm2	m	0.133	2.06	1.55	0.54	4.15
2.5mm2	m	0.133	2.06	1.84	0.59	4.49
4.0mm2	m	0.199	3.08	2.71	0.87	6.66
6.0mm2	m	0.199	3.08	3.42	0.98	7.48
10.0mm2	m	0.249	3.86	4.45	1.25	9.56
16.0mm2	m	0.249	3.86	6.21	1.51	11.58

	Unit	Hours	Hours £	Mat'ls £	O & P £	Total £
4 core cable 6944XL (cont'd)						
25.0mm2	m	0.349	5.41	3.32	1.31	10.04
35.0mm2	m	0.349	5.41	4.04	1.42	10.87

Complete outdoor (CW) termination comprising cable lugs, brass gland, locknut and earthing ring, PVC cable shroud including drilling and cutting steel gland plate and stripping 300mm long tails
NB Add 0.05 hours per metre for every extra 500mm of cable tails required

	Unit	Hours	Hours £	Mat'ls £	O & P £	Total £
2 core XLPE/SWA/LSF cable						
1.5mm2	each	0.483	7.49	11.98	2.92	22.39
2.5mm2	each	0.483	7.49	11.98	2.92	22.39
4.0mm2	each	0.483	7.49	12.12	2.94	22.55
6.0mm2	each	0.500	7.75	15.71	3.52	26.98
10.0mm2	each	0.650	10.08	15.82	3.88	29.78
16.0mm2	each	0.766	11.87	21.13	4.95	37.95
25.0mm2	each	0.983	15.24	21.16	5.46	41.86
35.0mm2	each	1.050	16.28	19.19	5.32	40.78
3 core XLPE/SWA/LSF cable						
1.5mm2	each	0.516	8.00	12.07	3.01	23.08
2.5mm2	each	0.516	8.00	12.07	3.01	23.08
4.0mm2	each	0.516	8.00	12.28	3.04	23.32
6.0mm2	each	0.533	8.26	15.87	3.62	27.75
10.0mm2	each	0.716	11.10	16.04	4.07	31.21
16.0mm2	each	0.833	12.91	21.32	5.13	39.37
25.0mm2	each	1.050	16.28	18.97	5.29	40.53
35.0mm2	each	1.150	17.83	19.45	5.59	42.87

	Unit	Hours	Hours £	Mat'ls £	O & P £	Total £
4 core XLPE/SWA/LSF cable						
1.5mm2	each	0.566	8.77	12.16	3.14	24.07
2.5mm2	each	0.566	8.77	12.16	3.14	24.07
4.0mm2	each	0.566	8.77	16.03	3.72	28.52
6.0mm2	each	0.583	9.04	16.03	3.76	28.83
10.0mm2	each	0.816	12.65	21.50	5.12	39.27
16.0mm2	each	0.933	14.46	21.54	5.40	41.40
25.0mm2	each	1.150	17.83	19.40	5.58	42.81
35.0mm2	each	1.300	20.15	20.05	6.03	46.23

Sundries

XLPE/SWA/LSF 'Telcleats'
LSF (low smoke and fume); to
backgrounds requiring fixings

	Unit	Hours	Hours £	Mat'ls £	O & P £	Total £
2 core cable to brickwork/ concrete						
1.5mm2	m	0.075	1.16	0.19	0.20	1.56
2.5mm2	m	0.075	1.16	0.19	0.20	1.56
4.0mm2	m	0.075	1.16	0.27	0.21	1.65
6.0mm2	m	0.075	1.16	0.32	0.22	1.70
10.0mm2	m	0.075	1.16	0.34	0.23	1.73
16.0mm2	m	0.075	1.16	0.34	0.23	1.73
25.0mm2	m	0.075	1.16	0.39	0.23	1.79
35.0mm2	m	0.083	1.29	0.39	0.25	1.93
3 core cable to brickwork/ concrete						
1.5mm2	m	0.075	1.16	0.19	0.20	1.56
2.5mm2	m	0.075	1.16	0.19	0.20	1.56
4.0mm2	m	0.075	1.16	0.27	0.21	1.65
6.0mm2	m	0.075	1.16	0.32	0.22	1.70
10.0mm2	m	0.075	1.16	0.34	0.23	1.73
16.0mm2	m	0.075	1.16	0.34	0.23	1.73
25.0mm2	m	0.083	1.29	0.39	0.25	1.93
35.0mm2	m	0.083	1.29	0.56	0.28	2.12

	Unit	Hours	Hours £	Mat'ls £	O & P £	Total £
4 core cable to brickwork/ concrete						
1.5mm2	m	0.075	1.16	0.19	0.20	1.56
2.5mm2	m	0.075	1.16	0.27	0.21	1.65
4.0mm2	m	0.075	1.16	0.32	0.22	1.70
6.0mm2	m	0.075	1.16	0.32	0.22	1.70
10.0mm2	m	0.075	1.16	0.34	0.23	1.73
16.0mm2	m	0.075	1.16	0.39	0.23	1.79
25.0mm2	m	0.083	1.29	0.56	0.28	2.12
35.0mm2	m	0.083	1.29	0.56	0.28	2.12
2 core cable to timber						
1.5mm2	m	0.058	0.90	0.19	0.16	1.25
2.5mm2	m	0.058	0.90	0.19	0.16	1.25
4.0mm2	m	0.058	0.90	0.27	0.18	1.34
6.0mm2	m	0.058	0.90	0.32	0.18	1.40
10.0mm2	m	0.058	0.90	0.34	0.19	1.42
16.0mm2	m	0.058	0.90	0.34	0.19	1.42
25.0mm2	m	0.058	0.90	0.39	0.19	1.48
35.0mm2	m	0.066	1.02	0.39	0.21	1.62
3 core cable to timber						
1.5mm2	m	0.058	0.90	0.19	0.16	1.25
2.5mm2	m	0.058	0.90	0.19	0.16	1.25
4.0mm2	m	0.058	0.90	0.27	0.18	1.34
6.0mm2	m	0.058	0.90	0.32	0.18	1.40
10.0mm2	m	0.058	0.90	0.34	0.19	1.42
16.0mm2	m	0.058	0.90	0.34	0.19	1.42
25.0mm2	m	0.066	1.02	0.39	0.21	1.62
35.0mm2	m	0.066	1.02	0.56	0.24	1.82
4 core cable to timber						
1.5mm2	m	0.058	0.90	0.19	0.16	1.25
2.5mm2	m	0.058	0.90	0.27	0.18	1.34
4.0mm2	m	0.058	0.90	0.32	0.18	1.40
6.0mm2	m	0.058	0.90	0.32	0.18	1.40
10.0mm2	m	0.058	0.90	0.34	0.19	1.42
16.0mm2	m	0.058	0.90	0.39	0.19	1.48
25.0mm2	m	0.066	1.02	0.56	0.24	1.82
35.0mm2	m	0.066	1.02	0.56	0.24	1.82

	Unit	Hours	Hours £	Mat'ls £	O & P £	Total £
Precast concrete warning tiles, laid in trenches						
cable covers						
914 × 152 × 63mm	m	0.116	1.80	5.12	1.04	7.96
914 × 229 × 63mm	m	0.125	1.94	5.01	1.04	7.99
914 × 305 × 63mm	m	0.133	2.06	7.17	1.38	10.62
Precast concrete markers, placed in excavation formed by others						
block type						
305 × 305 × 152mm	each	0.116	1.80	10.82	1.89	14.51
610 × 610 × 102mm	each	0.125	1.94	24.68	3.99	30.61
marker post						
pillar type	each	0.116	1.80	11.66	2.02	15.48
type 95 post	each	0.116	1.80	14.39	2.43	18.62
FP200 Gold low smoke zero halogen sheath (LS0H), stranded conductors; BS 5839 and BS 5266, to backgrounds requiring fixings (fixings measured elsewhere)						
2 core cable (red or white sheath)						
1.0mm2	m	0.108	1.67	1.71	0.51	3.89
1.5mm2	m	0.108	1.67	1.92	0.54	4.13
2.5mm2	m	0.125	1.94	2.66	0.69	5.29
4.0mm2	m	0.125	1.94	4.12	0.91	6.97
3 core cable						
1.0mm2	m	0.108	1.67	2.17	0.58	4.42
1.5mm2	m	0.108	1.67	2.72	0.66	5.05
2.5mm2	m	0.125	1.94	3.38	0.80	6.12

	Unit	Hours	Hours £	Mat'ls £	O & P £	Total £
4 core cable						
1.0mm2	m	0.116	1.80	2.65	0.67	5.12
1.5mm2	m	0.116	1.80	3.29	0.76	5.85
2.5mm2	m	0.141	2.19	4.61	1.02	7.81

Complete termination comprising nylon gland and locknut including drilling and cutting steel gland plate and stripping 150mm long tails (glands red or white finish)

	Unit	Hours	Hours £	Mat'ls £	O & P £	Total £
2 core FP200 Gold						
1.0mm2	each	0.133	2.06	0.44	0.38	2.88
1.5mm2	each	0.133	2.06	0.44	0.38	2.88
2.5mm2	each	0.133	2.06	0.44	0.38	2.88
4.0mm2	each	0.133	2.06	0.56	0.39	3.01
3 core FP200 Gold						
1.0mm2	each	0.166	2.57	0.44	0.45	3.46
1.5mm2	each	0.166	2.57	0.44	0.45	3.46
2.5mm2	each	0.166	2.57	0.56	0.47	3.60
4 core FP200 Gold						
1.0mm2	each	0.225	3.49	0.44	0.59	4.52
1.5mm2	each	0.225	3.49	0.44	0.59	4.52
2.5mm2	each	0.225	3.49	0.56	0.61	4.65

Sundries

FP200 Gold clips in red or white finish; to background requiring fixings (nails)

	Unit	Hours	Hours £	Mat'ls £	O & P £	Total £
2 core cable						
1.0mm2	m	0.016	0.25	0.08	0.05	0.38
1.5mm2	m	0.016	0.25	0.09	0.05	0.39
2.5mm2	m	0.016	0.25	0.10	0.05	0.40
4.0mm2	m	0.016	0.25	0.10	0.05	0.40

	Unit	Hours	Hours £	Mat'ls £	O & P £	Total £
3 core cable						
1.0mm2	m	0.016	0.25	0.08	0.05	0.38
1.5mm2	m	0.016	0.25	0.09	0.05	0.39
2.5mm2	m	0.016	0.25	0.09	0.05	0.39
4 core cable						
1.0mm2	m	0.016	0.25	0.09	0.05	0.39
1.5mm2	m	0.016	0.25	0.10	0.05	0.40
2.5mm2	m	0.016	0.25	0.10	0.05	0.40

500 volt grade, light duty mineral insulated LSF sheath copper conductors; to backgrounds requiring fixings (clips and fixings included in running length at 400mm average centres)

	Unit	Hours	Hours £	Mat'ls £	O & P £	Total £
2 core cable						
1.0mm2	m	0.107	1.66	2.72	0.66	5.04
1.5mm2	m	0.107	1.66	2.91	0.69	5.25
2.5mm2	m	0.107	1.66	3.58	0.79	6.02
4.0mm2	m	0.148	2.29	5.26	1.13	8.69
3 core cable						
1.0mm2	m	0.107	1.66	3.20	0.73	5.59
1.5mm2	m	0.107	1.66	3.83	0.82	6.31
2.5mm2	m	0.107	1.66	5.32	1.05	8.03
4 core cable						
1.0mm2	m	0.115	1.78	3.70	0.82	6.30
1.5mm2	m	0.115	1.78	4.58	0.95	7.32
2.5mm2	m	0.115	1.78	7.42	1.38	10.58
7 core cable						
1.5mm2	m	0.140	2.17	7.01	1.38	10.56
2.5mm2	m	0.165	2.56	8.98	1.73	13.27

	Unit	Hours	Hours £	Mat'ls £	O & P £	Total £
Complete termination comprising brass compression gland, light duty seal assemblies, screw-on pot, earth tail, brass locknut and LSF gland shroud including drilling and cutting steel gland plate and stripping 150mm long tails *NB Add 0.05 hours per metre for every extra 50mm of cable tails required*						
2 core cable MICV						
1.0mm2	m	0.350	5.43	3.50	1.34	10.26
1.5mm2	m	0.350	5.43	3.50	1.34	10.26
2.5mm2	m	0.350	5.43	3.50	1.34	10.26
4.0mm2	m	0.383	5.94	3.50	1.42	10.85
3 core cable MICV						
1.0mm2	m	0.383	5.94	3.50	1.42	10.85
1.5mm2	m	0.383	5.94	3.50	1.42	10.85
2.5mm2	m	0.383	5.94	3.50	1.42	10.85
4 core cable MICV						
1.0mm2	m	0.400	6.20	3.50	1.46	11.16
1.5mm2	m	0.400	6.20	3.50	1.46	11.16
2.5mm2	m	0.400	6.20	3.50	1.46	11.16
7 core cable MICV						
1.5mm2	m	0.530	8.22	10.47	2.80	21.49
2.5mm2	m	0.530	8.22	10.47	2.80	21.49

	Unit	Hours	Hours £	Mat'ls £	O & P £	Total £
750 volt grade, heavy duty mineral insulated LSF sheath copper conductors; to backgrounds requiring fixings (clips and fixings included in running length at 600mm average centres)						
single core cable						
10.0mm2	m	0.115	1.78	3.30	0.76	5.84
16.0mm2	m	0.115	1.78	4.52	0.95	7.25
25.0mm2	m	0.173	2.68	6.25	1.34	10.27
35.0mm2	m	0.173	2.68	8.19	1.63	12.50
50.0mm2	m	0.173	2.68	10.34	1.95	14.97
2 core cable						
1.5mm2	m	0.091	1.41	2.91	0.65	4.97
2.5mm2	m	0.091	1.41	3.45	0.73	5.59
4.0mm2	m	0.132	2.05	4.36	0.96	7.37
6.0mm2	m	0.132	2.05	5.78	1.17	9.00
10.0mm2	m	0.202	3.13	7.46	1.59	12.18
16.0mm2	m	0.202	3.13	10.54	2.05	15.72
25.0mm2	m	0.202	3.13	13.85	2.55	19.53
3 core cable						
1.5mm2	m	0.091	1.41	3.23	0.70	5.34
2.5mm2	m	0.091	1.41	3.96	0.81	6.18
4.0mm2	m	0.132	2.05	5.03	1.06	8.14
6.0mm2	m	0.132	2.05	6.33	1.26	9.63
10.0mm2	m	0.202	3.13	9.10	1.83	14.07
16.0mm2	m	0.202	3.13	12.84	2.40	18.37
25.0mm2	m	0.202	3.13	19.35	3.37	25.85
4 core cable						
1.5mm2	m	0.099	1.53	3.88	0.81	6.23
2.5mm2	m	0.099	1.53	4.88	0.96	7.38
4.0mm2	m	0.167	2.59	6.04	1.29	9.92
6.0mm2	m	0.167	2.59	7.94	1.58	12.11
10.0mm2	m	0.210	3.26	11.21	2.17	16.63
16.0mm2	m	0.210	3.26	16.35	2.94	22.55
25.0mm2	m	0.279	4.32	23.83	4.22	32.38

	Unit	Hours	Hours £	Mat'ls £	O & P £	Total £
Complete termination comprising brass compression gland, heavy duty seal assemblies, screw-on pot, earth tail, brass locknut and LSF gland shroud including drilling and cutting steel gland plate and stripping 150mm long tails *NB Add 0.05 hours per metre for every extra 50mm of cable tails required*						
single core cable MICV						
10.0mm2	each	0.216	3.35	6.58	1.49	11.42
16.0mm2	each	0.216	3.35	6.58	1.49	11.42
25.0mm2	each	0.300	4.65	6.58	1.68	12.91
35.0mm2	each	0.300	4.65	6.58	1.68	12.91
50.0mm2	each	0.300	4.65	7.80	1.87	14.32
2 core cable MICV						
1.5mm2	each	0.350	5.43	4.60	1.50	11.53
2.5mm2	each	0.350	5.43	4.60	1.50	11.53
4.0mm2	each	0.383	5.94	6.50	1.87	14.30
6.0mm2	each	0.383	5.94	6.50	1.87	14.30
10.0mm2	each	0.383	5.94	10.83	2.51	19.28
16.0mm2	each	0.416	6.45	18.18	3.69	28.32
25.0mm2	each	0.416	6.45	20.84	4.09	31.38
3 core cable MICV						
1.5mm2	each	0.383	5.94	4.66	1.59	12.19
2.5mm2	each	0.383	5.94	4.66	1.59	12.19
4.0mm2	each	0.416	6.45	7.25	2.05	15.75
6.0mm2	each	0.416	6.45	9.41	2.38	18.24
10.0mm2	each	0.416	6.45	12.92	2.91	22.27
16.0mm2	each	0.450	6.98	19.38	3.95	30.31
25.0mm2	each	0.450	6.98	32.07	5.86	44.90

	Unit	Hours	Hours £	Mat'ls £	O & P £	Total £
4 core cable MICV						
1.5mm2	each	0.400	6.20	4.31	1.58	12.09
2.5mm2	each	0.400	6.20	7.25	2.02	15.47
4.0mm2	each	0.433	6.71	8.82	2.33	17.86
6.0mm2	each	0.433	6.71	12.92	2.94	22.58
10.0mm2	each	0.433	6.71	12.93	2.95	22.59
16.0mm2	each	0.466	7.22	22.94	4.52	34.69
25.0mm2	each	0.466	7.22	32.13	5.90	45.26
450/750 volt grade PVC insulated cable, copper stranded conductors; BS 6004						
single core; drawn into conduit 6491X						
1.5mm2	m	0.025	0.39	0.34	0.11	0.84
2.5mm2	m	0.033	0.51	0.51	0.15	1.17
4.0mm2	m	0.033	0.51	1.14	0.25	1.90
6.0mm2	m	0.033	0.51	1.65	0.32	2.49
10.0mm2	m	0.041	0.64	3.09	0.56	4.28
16.0mm2	m	0.041	0.64	4.85	0.82	6.31
25.0mm2	m	0.058	0.90	4.57	0.82	6.29
35.0mm2	m	0.058	0.90	5.69	0.99	7.58
single core; drawn or laid into trunking 6491X						
1.5mm2	m	0.016	0.25	0.34	0.09	0.68
2.5mm2	m	0.016	0.25	0.51	0.11	0.87
4.0mm2	m	0.016	0.25	1.14	0.21	1.60
6.0mm2	m	0.016	0.25	1.65	0.28	2.18
10.0mm2	m	0.016	0.25	3.09	0.50	3.84
16.0mm2	m	0.016	0.25	4.85	0.76	5.86
25.0mm2	m	0.025	0.39	4.57	0.74	5.70
35.0mm2	m	0.025	0.39	5.69	0.91	6.99
50.0mm2	m	0.025	0.39	8.30	1.30	9.99

	Unit	Hours	Hours £	Mat'ls £	O & P £	Total £
300/500 volt grade, PVC insulated, PVC sheathed copper stranded conductors; BS 6004 to backgrounds requiring fixings (clips and fixings included in the running length at 150mm average centres)						
single core cable 6181Y						
1.0mm2	m	0.041	0.64	0.44	0.16	1.24
1.5mm2	m	0.041	0.64	0.58	0.18	1.40
2.5mm2	m	0.041	0.64	0.98	0.24	1.86
4.0mm2	m	0.041	0.64	1.60	0.34	2.57
6.0mm2	m	0.066	1.02	2.13	0.47	3.63
10.0mm2	m	0.066	1.02	3.38	0.66	5.06
16.0mm2	m	0.075	1.16	4.32	0.82	6.30
25.0mm2	m	0.075	1.16	8.28	1.42	10.86
35.0mm2	m	0.083	1.29	12.60	2.08	15.97
300/500 volt grade, PVC insulated, PVC sheathed twin and earth cabling; BS 6004 to backgrounds requiring fixings (clips and fixings included in the running length at 200mm average centres)						
twin and earth cabling 6242Y						
1.0mm2	m	0.058	0.90	0.56	0.22	1.68
1.5mm2	m	0.058	0.90	0.70	0.24	1.84
2.5mm2	m	0.058	0.90	0.96	0.28	2.14
4.0mm2	m	0.058	0.90	3.23	0.62	4.75
6.0mm2	m	0.075	1.16	3.84	0.75	5.75
10.0mm2	m	0.075	1.16	6.31	1.12	8.59
16.0mm2	m	0.100	1.55	10.06	1.74	13.35

	Unit	Hours	Hours £	Mat'ls £	O & P £	Total £
3 core and earth cable 6243Y						
1.0mm2	m	0.060	0.93	1.47	0.36	2.76
1.5mm2	m	0.060	0.93	2.31	0.49	3.73

300/500 volt grade, PVC insulated, PVC sheathed circular cables; copper stranded conductors; BS 6500; in tails including termination at both ends

	Unit	Hours	Hours £	Mat'ls £	O & P £	Total £
3 core cable 2183Y						
0.5mm2	m	0.133	2.06	0.37	0.36	2.80
0.75mm2	m	0.133	2.06	0.59	0.40	3.05

300/500 volt grade, PVC insulated, PVC sheathed heat resistant cables circular cables; copper stranded conductors; BS 6141; in tails including termination at both ends

	Unit	Hours	Hours £	Mat'ls £	O & P £	Total £
3 core cables 3093Y						
0.5mm2	m	0.133	2.06	0.81	0.43	3.30
0.75mm2	m	0.133	2.06	0.87	0.44	3.37
1.0mm2	m	0.133	2.06	1.18	0.49	3.73
1.5mm2	m	0.150	2.33	1.64	0.59	4.56
2.5mm2	m	0.158	2.45	2.46	0.74	5.65

	Unit	Hours	Hours £	Mat'ls £	O & P £	Total £
Y71 LV SWITCHGEAR AND DISTRUBUTION BOARDS						
Consumer units as manufactured by Crabtree in their 'Starbreaker' range (modular concept)						
Insulated surface mounted consumer unit carcase less main switch; to backgrounds requiring fixings						
4-way module	each	0.600	9.30	6.01	2.30	17.61
6-way module	each	0.600	9.30	10.24	2.93	22.47
9-way module	each	0.650	10.08	13.41	3.52	27.01
12-way module	each	0.650	10.08	16.22	3.94	30.24
15-way module	each	0.766	11.87	21.63	5.03	38.53
17-way module	each	0.882	13.67	30.44	6.62	50.73
20-way module	each	0.882	13.67	29.01	6.40	49.08
Metal cased surface mounted consumer unit carcase less main switch; to backgrounds requiring fixings						
8-way module	each	0.650	10.08	15.52	3.84	29.43
10-way module	each	0.650	10.08	18.85	4.34	33.26
14-way module	each	0.766	11.87	25.09	5.54	42.51
17-way module	each	0.882	13.67	25.30	5.85	44.82
20-way module	each	0.882	13.67	37.49	7.67	58.84
28-way module	each	0.998	15.47	44.82	9.04	69.33
Consumer unit accessories; fix/ place on to DIN rail						
Incoming devices						
100 amp DP switch disconnector	each	0.250	3.88	9.58	2.02	15.47
100 amp DP switch disconnector complete with tap-off terminal	each	0.250	3.88	10.64	2.18	16.69

	Unit	Hours	Hours £	Mat'ls £	O & P £	Total £
Consumer unit accessories (cont'd)						
100 amp SP direct connection unit	each	0.250	3.88	4.89	1.31	10.08
25 amp; 30 milli amp main incoming RCCB	each	0.250	3.88	44.25	7.22	55.34
80 amp; 30 milli amp main incoming RCCB	each	0.250	3.88	55.17	8.86	67.90
80 amp; 30 milli amp main incoming RCCB Type A	each	0.500	7.75	63.75	10.73	82.23
80 amp; 100 milli amp main incoming RCCB	each	0.500	7.75	53.07	9.12	69.94
63 amp; 30 milli amp split load RCCB	each	0.500	7.75	57.26	9.75	74.76
63 amp; 100 milli amp split load RCCB	each	0.500	7.75	53.23	9.15	70.13
80 amp; 30 milli amp split load RCCB	each	0.500	7.75	59.85	10.14	77.74
80 amp; 100 milli amp split load RCCB	each	0.500	7.75	55.63	9.51	72.89
Miniature circuit breakers (MCBs)						
6 amp; single pole	each	0.150	2.33	6.85	1.38	10.55
10 amp; single pole	each	0.150	2.33	6.47	1.32	10.11
16 amp; single pole	each	0.150	2.33	6.47	1.32	10.11
20 amp; single pole	each	0.150	2.33	6.47	1.32	10.11
32 amp; single pole	each	0.150	2.33	6.47	1.32	10.11
40 amp; single pole	each	0.150	2.33	6.85	1.38	10.55
6 amp; Type 1 single pole	each	0.150	2.33	7.71	1.51	11.54
10 amp; Type 1 single pole	each	0.150	2.33	7.29	1.44	11.06
16 amp; Type 1 single pole	each	0.150	2.33	7.29	1.44	11.06
20 amp; Type 1 single pole	each	0.150	2.33	7.29	1.44	11.06
32 amp; Type 1 single pole	each	0.150	2.33	7.29	1.44	11.06
40 amp; Type 1 single pole	each	0.150	2.33	7.71	1.51	11.54

	Unit	Hours	Hours £	Mat'ls £	O & P £	Total £
Residual current circuit breakers (RCCBs); double pole						
40 amp; 30 milli amp	each	0.500	7.75	37.67	6.81	52.23
63 amp; 30 milli amp	each	0.500	7.75	45.29	7.96	61.00
80 amp; 30 milli amp	each	0.500	7.75	53.95	9.26	70.96
80 amp; 100 milli amp	each	0.500	7.75	46.00	8.06	61.81
100 amp; 30 milli amp	each	0.500	7.75	64.20	10.79	82.74
Miniature circuit breakers/residual current circuit breakers (MCB/ RCCBs)						
6 amp; 30 milli amp	each	0.300	4.65	46.62	7.69	58.96
16 amp; 10 milli amp	each	0.300	4.65	51.15	8.37	64.17
16 amp; 30 milli amp	each	0.300	4.65	46.62	7.69	58.96
20 amp; 30 milli amp	each	0.300	4.65	46.62	7.69	58.96
32 amp; 10 milli amp	each	0.300	4.65	50.67	8.30	63.62
32 amp; 30 milli amp	each	0.300	4.65	46.19	7.63	58.47
32 amp; 30 milli amp with over current protection	each	0.300	4.65	53.12	8.67	66.44
40 amp; 30 milli amp	each	0.300	4.65	46.62	7.69	58.96
Additional accessories						
lighting time switch, 1 to 7 minutes, adjustable	each	0.166	2.57	27.12	4.45	34.15
bell transformer, 4-8-12 volt	each	0.166	2.57	17.82	3.06	23.45
analogue quartz timer - 7 day	each	0.166	2.57	63.32	9.88	75.78
blank plate, 1 way	each	0.033	0.51	0.58	0.16	1.26
brass padlock and keys	each	0.025	0.39	3.58	0.60	4.56
terminal cover	each	0.016	0.25	2.80	0.46	3.51

	Unit	Hours	Hours £	Mat'ls £	O & P £	Total £
Consumer unit busbar links; fixed into DIN rail accessories						
busbars						
6 module/4 MCB ways	each	0.266	4.12	2.27	0.96	7.35
8 module/6 MCB ways	each	0.266	4.12	2.52	1.00	7.64
10 module/8 MCB ways	each	0.299	4.63	3.46	1.21	9.31
14 module/12 MCB ways	each	0.365	5.66	4.97	1.59	12.22
17 module/15 MCB ways	each	0.398	6.17	5.20	1.71	13.07
20 module/18 MCB ways	each	0.431	6.68	6.83	2.03	15.54
split load busbars						
9 module/2 + 2 MCB ways	each	0.299	4.63	4.83	1.42	10.88
10 module/2 + 3 MCB ways	each	0.299	4.63	4.83	1.42	10.88
12 module/2 + 5 MCB ways	each	0.332	5.15	5.90	1.66	12.70
14 module/3 + 6 MCB ways	each	0.365	5.66	5.90	1.73	13.29
14 module/4 + 5 MCB ways	each	0.365	5.66	5.90	1.73	13.29
14 module/5 + 4 MCB ways	each	0.365	5.66	5.90	1.73	13.29
14 module/6 + 3 MCB ways	each	0.365	5.66	5.90	1.73	13.29
20 module/5 + 10 MCB way	each	0.431	6.68	8.06	2.21	16.95
20 module/8 + 7 MCB ways	each	0.431	6.68	8.06	2.21	16.95
20 module/6 + 9 MCB ways	each	0.431	6.68	8.06	2.21	16.95
dual tarriff 20 module/3 + 6 + 4 MCB ways	each	0.431	6.68	8.06	2.21	16.95
17 module/6 + 6 MCB ways	each	0.332	5.15	6.32	1.72	13.19
17 module/5 + 7 MCB ways	each	0.332	5.15	6.32	1.72	13.19
17 module/7 + 5 MCB ways	each	0.332	5.15	6.32	1.72	13.19
17 module/ 9 + 3 MCB ways	each	0.332	5.15	6.32	1.72	13.19

Distribution boards as manufactured by Bill in their 'Talisman Plus' range

	Unit	Hours	Hours £	Mat'ls £	O & P £	Total £
Mild-steel surface mounted distribution boards including integral incoming switch; to backgrounds requiring fixings						
4 way; SP&N; 100 amp main switch	each	0.366	5.67	36.05	6.26	47.98

	Unit	Hours	Hours £	Mat'ls £	O & P £	Total £
7 way; SP&N; 100 amp main switch	each	0.450	6.98	44.39	7.70	59.07
10 way; SP&N; 100 amp main switch	each	0.500	7.75	52.77	9.08	69.60
13 way; SP&N; 100 amp main switch	each	0.550	8.53	60.54	10.36	79.42
16 way; SP&N; 100 amp main switch	each	0.683	10.59	71.86	12.37	94.81
4 way; TP&N; 100 amp main switch	each	0.533	8.26	129.33	20.64	158.23
6 way; TP&N; 100 amp main switch	each	0.800	12.40	140.45	22.93	175.78
8 way; TP&N; 100 amp main switch	each	0.866	13.42	156.11	25.43	194.96
12 way; TP&N; 100 amp main switch	each	1.000	15.50	178.16	29.05	222.71
16 way; TP&N; 100 amp main switch	each	1.083	16.79	230.68	37.12	284.59

Distribution boards miniature circuit breakers (MCBs); fix/place on to DIN rail

	Unit	Hours	Hours £	Mat'ls £	O & P £	Total £
6 amp; single pole	each	0.150	2.33	6.38	1.31	10.01
10 amp; single pole	each	0.150	2.33	6.04	1.25	9.62
16 amp; single pole	each	0.150	2.33	6.04	1.25	9.62
20 amp; single pole	each	0.150	2.33	6.04	1.25	9.62
25 amp; single pole	each	0.150	2.33	6.04	1.25	9.62
32 amp; single pole	each	0.150	2.33	6.04	1.25	9.62
40 amp; single pole	each	0.150	2.33	6.38	1.31	10.01
50 amp; single pole	each	0.150	2.33	6.38	1.31	10.01
63 amp; single pole	each	0.166	2.57	6.38	1.34	10.30
6 amp; double pole	each	0.300	4.65	18.15	3.42	26.22
10 amp; double pole	each	0.300	4.65	17.29	3.29	25.23
16 amp; double pole	each	0.300	4.65	17.29	3.29	25.23
6 amp; triple pole	each	0.450	6.98	26.84	5.07	38.89
10 amp; triple pole	each	0.450	6.98	25.61	4.89	37.47
16 amp; triple pole	each	0.450	6.98	25.61	4.89	37.47
20 amp; triple pole	each	0.450	6.98	25.61	4.89	37.47
25 amp; triple pole	each	0.450	6.98	25.61	4.89	37.47

	Unit	Hours	Hours £	Mat'ls £	O & P £	Total £
32 amp; triple pole	each	0.450	6.98	25.61	4.89	37.47
40 amp; triple pole	each	0.450	6.98	26.84	5.07	38.89
50 amp; triple pole	each	0.450	6.98	26.84	5.07	38.89
63 amp; triple pole	each	0.500	7.75	26.84	5.19	39.78
Residual current breaker operators (RCBOs); fix/place on to DIN rail						
6 amp; 30 milli amp; single pole	each	0.150	2.33	42.25	6.69	51.26
10 amp; 30 milli amp; single pole	each	0.150	2.33	41.92	6.64	50.88
16 amp; 30 milli amp; single pole	each	0.150	2.33	41.92	6.64	50.88
20 amp; 30 milli amp; single pole	each	0.150	2.33	41.92	6.64	50.88
32 amp; 30 milli amp; single pole	each	0.150	2.33	41.92	6.64	50.88
40 amp; 30 milli amp; single pole	each	0.150	2.33	42.25	6.69	51.26
Sundry items of distribution board equipment						
Fix/place on to DIN rail						
single pole blank plate	each	0.033	0.51	0.52	0.15	1.19
To backgrounds requiring fixings						
single pole board locking kit	each	0.200	3.10	5.96	1.36	10.42
triple pole and neutral board locking kit	each	0.200	3.10	5.96	1.36	10.42

	Unit	Hours	Hours £	Mat'ls £	O & P £	Total £
Y73 LUMINAIRES AND LAMPS						
Recessed modular sheet steel powder-coated paint finish, enclosed box construction, fluorescent luminaires complete control gear and lamps; to backgrounds requiring fixings (diffusers measured elsewhere)						
Switch start						
600 × 600mm; 3 lamp; 18 watt	each	1.250	19.38	18.25	5.64	43.27
1200 × 600mm; 3 lamp; 36 watt	each	1.800	27.90	24.40	7.85	60.15
600 × 600mm; 4 lamp; 18 watt	each	1.500	23.25	20.35	6.54	50.14
1200 × 600mm; 4 lamp; 36 watt	each	2.025	31.39	26.65	8.71	66.74
High frequency						
600 × 600mm; 3 lamp; 18 watt	each	1.250	19.38	42.75	9.32	71.44
1200 × 600mm; 3 lamp; 36 watt	each	1.800	27.90	65.90	14.07	107.87
600 × 600mm; 4 lamp; 18 watt	each	1.500	23.25	46.35	10.44	80.04
1200 × 600mm; 4 lamp; 36 watt	each	2.025	31.39	70.55	15.29	117.23
Recessed modular sheet steel powder-coated paint finish, enclosed box construction, fluorescent luminaires complete integral emergency conversion unit (3 hour maintained module and battery pack to maintain one lamp in the event of mains failure) control gear and lamps; to backgrounds requiring fixings (diffusers measured elsewhere)						

	Unit	Hours	Hours £	Mat'ls £	O & P £	Total £
Switch start						
600 × 600mm; 3 lamp; 18 watt	each	1.250	19.38	115.45	20.22	155.05
1200 × 600mm; 3 lamp; 36 watt	each	1.800	27.90	112.95	21.13	161.98
600 × 600mm; 4 lamp; 18 watt	each	1.500	23.25	110.95	20.13	154.33
1200 × 600mm; 4 lamp; 36 watt	each	2.025	31.39	114.55	21.89	167.83
High frequency						
600 × 600mm; 3 lamp; 18 watt	each	1.250	19.38	136.20	23.34	178.91
1200 × 600mm; 3 lamp; 36 watt	each	1.800	27.90	155.58	27.52	211.00
600 × 600mm; 4 lamp; 18 watt	each	1.500	23.25	140.95	24.63	188.83
1200 × 600mm; 4 lamp; 36 watt	each	2.025	31.39	164.95	29.45	225.79
Diffusers suitable for recessed and surface modular luminaires; insert into luminaire body						
Category 2 louvres						
600 × 600mm; 3 lamp; 18 watt	each	0.166	2.57	19.95	3.38	25.90
1200 × 600mm; 3 lamp; 36 watt	each	0.166	2.57	28.95	4.73	36.25
600 × 600mm; 4 lamp; 18 watt	each	0.166	2.57	20.95	3.53	27.05
1200 × 600mm; 4 lamp; 36 watt	each	0.166	2.57	30.95	5.03	38.55
low brightness louvres						
600 × 600mm; 3 lamp; 18 watt	each	0.166	2.57	12.95	2.33	17.85
1200 × 600mm; 3 lamp; 36 watt	each	0.166	2.57	20.95	3.53	27.05

	Unit	Hours	Hours £	Mat'ls £	O & P £	Total £
low brightness louvres (cont'd)						
600 × 600mm; 4 lamp; 18 watt	each	0.166	2.57	13.95	2.48	19.00
1200 × 600mm; 4 lamp; 36 watt	each	0.166	2.57	21.95	3.68	28.20
clear styrene flat sheet diffuser						
600 × 600mm; 3 lamp; 18 watt	each	0.133	2.06	1.95	0.60	4.61
1200 × 600mm; 3 lamp; 36 watt	each	0.133	2.06	3.95	0.90	6.91
600 × 600mm; 4 lamp; 18 watt	each	0.133	2.06	1.95	0.60	4.61
1200 × 600mm; 4 lamp; 36 watt	each	0.133	2.06	3.95	0.90	6.91

Surface-mounted fluorescent luminaires, sheet steel body, powder-coated paint finish with colour matched moulded end caps complete with control gear, lamps and louvre; to backgrounds requiring fixings

	Unit	Hours	Hours £	Mat'ls £	O & P £	Total £
Switch start, category 2 louvres						
1253 × 184 × 83mm (4 ft; 36 watt single tube)	each	1.116	17.30	46.70	9.60	73.60
1553 × 184 × 83mm (5 ft; 58 watt single tube)	each	1.150	17.83	52.45	10.54	80.82
1817 × 184 × 83mm (6 ft; 70 watt single tube)	each	1.300	20.15	62.95	12.47	95.57

	Unit	Hours	Hours £	Mat'ls £	O & P £	Total £
Switch start, category 2 louvres (cont'd)						
1253 × 280 × 83mm (4 ft; 36 watt twin tube)	each	1.383	21.44	58.45	11.98	91.87
1553 × 280 × 83mm (5 ft; 58 watt twin tube)	each	1.416	21.95	69.95	13.78	105.68
1817 × 280 × 83mm (6 ft; 70 watt twin tube)	each	1.582	24.52	80.95	15.82	121.29
643 × 510 × 83mm (2 ft; 18 watt four tube)	each	0.766	11.87	78.95	13.62	104.45
Switch start, low brightness louvres						
1253 × 184 × 83mm (4 ft; 36 watt single tube)	each	1.116	17.30	43.25	9.08	69.63
1553 × 184 × 83mm (5 ft; 58 watt single tube)	each	1.150	17.83	51.00	10.32	79.15
1817 × 184 × 83mm (6 ft; 70 watt single tube)	each	1.300	20.15	59.50	11.95	91.60
1253 × 280 × 83mm (4 ft; 36 watt twin tube)	each	1.383	21.44	55.00	11.47	87.90
1553 × 280 × 83mm (5 ft; 58 watt twin tube)	each	1.416	21.95	66.50	13.27	101.72
1817 × 280 × 83mm (6 ft; 70 watt twin tube)	each	1.582	24.52	77.50	15.30	117.32
643 × 510 × 83mm (2 ft; 18 watt four tube)	each	0.766	11.87	75.50	13.11	100.48
High frequency; category 2 louvres						
1253 × 184 × 83mm (4 ft; 36 watt single tube)	each	1.116	17.30	66.70	12.60	96.60
1553 × 184 × 83mm (5 ft; 58 watt single tube)	each	1.150	17.83	76.45	14.14	108.42
1817 × 184 × 83mm (6 ft; 70 watt single tube)	each	1.300	20.15	84.95	15.77	120.87
1253 × 280 × 83mm (4 ft; 36 watt twin tube)	each	1.383	21.44	77.45	14.83	113.72

	Unit	Hours	Hours £	Mat'ls £	O & P £	Total £
High frequency; category 2 louvres (cont'd)						
1553 × 280 × 83mm (5 ft; 58 watt twin tube)	each	1.416	21.95	85.45	16.11	123.51
1817 × 280 × 83mm (6 ft; 70 watt twin tube)	each	1.582	24.52	96.96	18.22	139.70
643 × 510 × 83mm (2 ft; 18 watt four tube)	each	0.766	11.87	94.95	16.02	122.85
High frequency, low brightness louvres						
1253 × 184 × 83mm (4 ft; 36 watt single tube)	each	1.116	17.30	60.25	11.63	89.18
1553 × 184 × 83mm (5 ft; 58 watt single tube)	each	1.150	17.83	72.00	13.47	103.30
1817 × 184 × 83mm (6 ft; 70 watt single tube)	each	1.300	20.15	81.50	15.25	116.90
1253 × 280 × 83mm (4 ft; 36 watt twin tube)	each	1.383	21.44	74.00	14.32	109.75
1553 × 280 × 83mm (5 ft; 58 watt twin tube)	each	1.416	21.95	85.50	16.12	123.57
1817 × 280 × 83mm (6 ft; 70 watt twin tube)	each	1.582	24.52	94.50	17.85	136.87
643 × 510 × 83mm (2 ft; 18 watt four tube)	each	0.766	11.87	91.50	15.51	118.88

Surface-mounted fluorescent luminaire, sheet steel body powder-coated paint finish, with colour matched moulded end caps complete with intergral emergency conversion unit (3 hour maintained module and battery pack to maintain one lamp in the event of mains failure) control gear and lamps and louvre; to backgrounds requiring fixings

	Unit	Hours	Hours £	Mat'ls £	O & P £	Total £
Switch start, category 2 louvres						
1253 × 184 × 83mm (4 ft; 36 watt single tube)	each	1.116	17.30	100.70	17.70	135.70
1553 × 184 × 83mm (5 ft; 58 watt single tube)	each	1.150	17.83	112.45	19.54	149.82
1817 × 184 × 83mm (6 ft; 70 watt single tube)	each	1.300	20.15	123.95	21.62	165.72
1253 × 280 × 83mm (4 ft; 36 watt twin tube)	each	1.383	21.44	111.45	19.93	152.82
1553 × 280 × 83mm (5 ft; 58 watt twin tube)	each	1.416	21.95	123.95	21.88	167.78
1817 × 280 × 83mm (6 ft; 70 watt twin tube)	each	1.582	24.52	138.95	24.52	187.99
643 × 510 × 83mm (2 ft; 18 watt four tube)	each	0.766	11.87	126.95	20.82	159.65
Switch start, low brightness louvres						
1253 × 184 × 83mm (4 ft; 36 watt single tube)	each	1.116	17.30	97.25	17.18	131.73
1553 × 184 × 83mm (5 ft; 58 watt single tube)	each	1.150	17.83	109.00	19.02	145.85
1817 × 184 × 83mm (6 ft; 70 watt single tube)	each	1.300	20.15	120.50	21.10	161.75
1253 × 280 × 83mm (4 ft; 36 watt twin tube)	each	1.383	21.44	108.00	19.42	148.85
1553 × 280 × 83mm (5 ft; 58 watt twin tube)	each	1.416	21.95	120.50	21.37	163.82
1817 × 280 × 83mm (6 ft; 70 watt twin tube)	each	1.582	24.52	135.50	24.00	184.02
643 × 510 × 83mm (2 ft; 18 watt four tube)	each	0.766	11.87	123.50	20.31	155.68
High frequency; category 2 louvres						
1253 × 184 × 83mm (4 ft; 36 watt single tube)	each	1.116	17.30	115.70	19.95	152.95
1553 × 184 × 83mm (5 ft; 58 watt single tube)	each	1.150	17.83	127.45	21.79	167.07

	Unit	Hours	Hours £	Mat'ls £	O & P £	Total £
1817 × 184 × 83mm (6 ft; 70 watt single tube)	each	1.300	20.15	139.95	24.02	184.12
1253 × 280 × 83mm (4 ft; 36 watt twin tube)	each	1.383	21.44	126.46	22.18	170.08
1553 × 280 × 83mm (5 ft; 58 watt twin tube)	each	1.416	21.95	138.95	24.13	185.03
1817 × 280 × 83mm (6 ft; 70 watt twin tube)	each	1.582	24.52	154.95	26.92	206.39
643 × 510 × 83mm (2 ft; 18 watt four tube)	each	0.766	11.87	144.95	23.52	180.35
High frequency, low brightness louvres						
1253 × 184 × 83mm (4 ft; 36 watt single tube)	each	1.116	17.30	112.25	19.43	148.98
1553 × 184 × 83mm (5 ft; 58 watt single tube)	each	1.150	17.83	124.00	21.27	163.10
1817 × 184 × 83mm (6 ft; 70 watt single tube)	each	1.300	20.15	136.50	23.50	180.15
1253 × 280 × 83mm (4 ft; 36 watt twin tube)	each	1.383	21.44	123.00	21.67	166.10
1553 × 280 × 83mm (5 ft; 58 watt twin tube)	each	1.416	21.95	135.50	23.62	181.07
1817 × 280 × 83mm (6 ft; 70 watt twin tube)	each	1.582	24.52	151.50	26.40	202.42
643 × 510 × 83mm (2 ft; 18 watt four tube)	each	0.766	11.87	141.50	23.01	176.38
Surface-mounted batten fluorescent luminaires complete with control gear, lamp and opal diffuser; to backgrounds requiring fixings						
Popular pack (manufactured by Thorn Lighting)						
2 ft; single tube	each	0.633	9.81	6.35	2.42	18.59
4 ft; single tube	each	0.950	14.73	7.20	3.29	25.21

	Unit	Hours	Hours £	Mat'ls £	O & P £	Total £
Popular pack (cont'd)						
5 ft; single tube	each	0.950	14.73	8.55	3.49	26.77
6 ft; single tube	each	1.100	17.05	10.80	4.18	32.03
8 ft; single tube	each	1.250	19.38	20.96	6.05	46.39
2 ft; twin tube	each	0.658	10.20	9.67	2.98	22.85
4 ft; twin tube	each	0.975	15.11	12.60	4.16	31.87
5 ft; twin tube	each	0.975	15.11	16.20	4.70	36.01
6 ft; twin tube	each	1.125	17.44	21.60	5.86	44.89
8 ft; twin tube	each	1.275	19.76	38.47	8.73	66.97
Surface mini-line mounted batten fluorescent luminaires complete with control gear, lamp and diffuser; to backgrounds requiring fixings						
Mini-line connect (manufactured by Concorde)						
6 watt; T5 261mm long	each	0.633	9.81	16.60	3.96	30.37
8 watt; T5 337mm long	each	0.633	9.81	16.60	3.96	30.37
13 watt; T5 566mm long	each	0.633	9.81	17.94	4.16	31.91
15 watt; T8 490mm long	each	0.633	9.81	17.94	4.16	31.91
Mini-line accessories; plug-in/ connect; to backgrounds requiring fi×ings						
cables and plugs						
main lead; 1000mm long	each	0.250	3.88	4.75	1.29	9.92
connector lead; 500mm long and two plugs	each	0.033	0.51	5.70	0.93	7.14
end blanking plugs	each	0.008	0.12	0.38	0.08	0.58

	Unit	Hours	Hours £	Mat'ls £	O & P £	Total £
Surface-mounted dual voltage shaver light to BS EN 60598 and EN 60742 IP21 rated complete with lamp and pull switch; to backgrounds requiring fixings						
Shaver light						
11 watt; 425 × 56 × 70mm	each	0.950	14.73	11.69	3.96	30.38
Surface-mounted tungsten luminaire picture light finished in brass complete with strip light; to backgrounds requiring fixings						
Picture light						
60 watt; 130 × 280 × 200mm	each	0.616	9.55	23.47	4.95	37.97
60 watt; 150 × 285 × 200mm	each	0.616	9.55	25.37	5.24	40.16
Recessed display and undershelf low voltage downlighter kit complete with dichroic reflector lamps (3 nr), collars (3 nr), diffusers for surface mounting (3 nr), transformer (1 nr) and wiring loom, to backgrounds requiring fixings						
Low voltage downlighter kit						
3 × 20 watt; 12 volt halogen	each	0.850	13.18	26.96	6.02	46.16

	Unit	Hours	Hours £	Mat'ls £	O & P £	Total £
Recessed low voltage eyeball downlighter kit complete with dichroic reflector lamps (3 nr), transformer (1 nr) and interconnecting cables (2 m); into preformed recess formed by others						
Low voltage eyeball downlighter kit						
3 × 20 watt; 12 volt halogen	each	0.850	13.18	19.95	4.97	38.09
Recessed compact fluorescent downlighter complete with two pin compact fluorescent lamp; into preformed recess formed by others						
Recessed downlighter; 240 volt						
10 watt; fluorescent	each	0.270	4.19	22.52	4.01	30.71
2 × 10 watt; fluorescent	each	0.290	4.50	29.64	5.12	39.26
Surface low voltage track complete with spot light kit, transformer and halogen lamps; to backgrounds requiring fixings						
Low voltage track and spot kit						
3 × 20 watt; 12 volt halogen	each	0.599	9.28	37.95	7.09	54.32
Recessed mains voltage shower lights manufactured to BS LN 60598 suitable for zones 1 to 3 bathroom use complete with opal lens and lamp; into preformed recess by others						

	Unit	Hours	Hours £	Mat'ls £	O & P £	Total £
Recessed tungsten reflector downlighter	each	0.270	4.19	11.95	2.42	18.56
Surface mounted ceiling mounted luminair suitable for zones 1 to 3 bathroom use complete with diffuser and GLS lamp; to backgrounds requiring fixings						
Surface tungsten ceiling luminaire						
60 watt; chrome dome	each	0.800	12.40	9.71	3.32	25.43
Surface-mounted polycarbonate construction vandal resistant, 3 hour duration, emergency luminaire IP65 complete with sealed nickel cadmium batteries and lamp; to backgrounds requiring fixings						
Non-maintained						
8 watt; bulkhead	each	0.900	13.95	21.56	5.33	40.84
Maintained						
8 watt; bulkhead	each	0.900	13.95	35.96	7.49	57.40
Surface-mounted sheet steel white epoxy powder-coated emergency exit luminaire complete with sealed nickel cadmium batteries, 3 hour duration, hinged removable gear tray, lamp and EXIT legend sign (arrow); to backgrounds requiring fixings						

	Unit	Hours	Hours £	Mat'ls £	O & P £	Total £
Non-maintained						
8 watt; 30m viewing distance	each	0.900	13.95	49.41	9.50	72.86
Maintained						
8 watt; 30m viewing distance	each	0.900	13.95	58.41	10.85	83.21
Surface mounted external bulkhead luminaire as manufactured by Coughtrie and complete with opal diffuser and 100 watt lamp						
Coughtrie external bulkhead luminaire; to backgrounds requiring fixings	each	0.750	11.63	23.65	5.29	40.57
Surface-mounted sheet steel white and black epoxy powder-coated box complete with 12 volt sealed lead acid batteries, 3 hour duration and twin floodlight lamp set; place in position						
Non-maintained						
12 volt; 21 watt unit	each	0.650	10.08	107.96	17.71	135.74
Tungsten pendant set comprising ceiling rose, 225mm long white flexible cable and lamp holder (excludes lamp); to backgrounds requiring fixings						
Pendant set	each	0.233	3.61	3.08	1.00	7.70

	Unit	Hours	Hours £	Mat'ls £	O & P £	Total £
Y74 ACCESSORIES FOR ELECTRICAL SERVICES						
250 volt grade flush mounted accessories; including metal back boxes; to backgrounds requiring fixings						
White plastic switch plates; 6 amp rated						
1 gang; 1 way; single pole	each	0.216	3.35	1.44	0.72	5.51
1 gang; 2 way; single pole	each	0.249	3.86	1.61	0.82	6.29
2 gang; 2 way; single pole	each	0.348	5.39	2.40	1.17	8.96
3 gang; 2 way; single pole	each	0.448	6.94	6.35	1.99	15.29
4 gang; 2 way; single pole	each	0.581	9.01	10.82	2.97	22.80
6 gang; 2 way; single pole	each	0.880	13.64	21.20	5.23	40.07
1 gang; intermediate; single pole	each	0.282	4.37	6.05	1.56	11.98
1 gang; 1 way; double pole	each	0.282	4.37	4.66	1.35	10.39
1 gang; marked with 'Bell' symbol	each	0.216	3.35	3.75	1.06	8.16
1 gang; marked 'Press'	each	0.216	3.35	3.75	1.06	8.16
White plastic architrave switches; 6 amp rated						
1 gang; 1 way; single pole	each	0.216	3.35	3.08	0.96	7.39
1 gang; 2 way; single pole	each	0.249	3.86	3.38	1.09	8.33
2 gang; 2 way; single pole	each	0.365	5.66	6.83	1.87	14.36
1 gang; marked with 'Bell' symbol	each	0.216	3.35	4.21	1.13	8.69
1 gang; marked 'Press'	each	0.216	3.35	4.21	1.13	8.69
250 volt grade flush mounted accessories; including dry lining back boxes; fixed into apertures						

	Unit	Hours	Hours £	Mat'ls £	O & P £	Total £
White plastic switch plates; 6 amp rated						
1 gang; 1 way; single pole	each	0.183	2.84	1.39	0.63	4.86
1 gang; 2 way; single pole	each	0.216	3.35	1.55	0.73	5.63
2 gang; 2 way; single pole	each	0.249	3.86	2.30	0.92	7.08
3 gang; 2 way; single pole	each	0.415	6.43	6.21	1.90	14.54
4 gang; 2 way; single pole	each	0.548	8.49	11.09	2.94	22.52
1 gang; intermediate; single pole	each	0.249	3.86	6.16	1.50	11.52
1 gang; 1 way; double pole	each	0.249	3.86	4.77	1.29	9.92
1 gang; marked with 'Bell' symbol	each	0.183	2.84	3.72	0.98	7.54
1 gang; marked 'Press'	each	0.183	2.84	3.72	0.98	7.54

250 volt grade surface-mounted accessories; including moulded plastic back boxes; to backgrounds requiring fixings

	Unit	Hours	Hours £	Mat'ls £	O & P £	Total £
White plastic switch plates; 6 amp rated						
1 gang; 1 way; single pole	each	0.250	3.88	1.82	0.85	6.55
1 gang; 2 way; single pole	each	0.283	4.39	2.38	1.01	7.78
2 gang; 2 way; single pole	each	0.382	5.92	3.46	1.41	10.79
3 gang; 2 way; single pole	each	0.482	7.47	6.79	2.14	16.40
4 gang; 2 way; single pole	each	0.681	10.56	11.53	3.31	25.40
6 gang; 2 way; single pole	each	0.946	14.66	20.94	5.34	40.94
1 gang; intermediate; single pole	each	0.316	4.90	6.18	1.66	12.74
1 gang; 1 way; double pole	each	0.316	4.90	5.21	1.52	11.62
1 gang; marked with 'Bell' symbol	each	0.250	3.88	4.19	1.21	9.27
1 gang; marked 'Press'	each	0.250	3.88	4.19	1.21	9.27

	Unit	Hours	Hours £	Mat'ls £	O & P £	Total £
White plastic architrave switches; 6 amp rated						
1 gang; 1 way; single pole	each	0.216	3.35	3.31	1.00	7.66
1 gang; 2 way; single pole	each	0.249	3.86	3.63	1.12	8.61
2 gang; 2 way; single pole	each	0.398	6.17	7.34	2.03	15.54
1 gang; marked with 'Bell' symbol	each	0.216	3.35	4.53	1.18	9.06
1 gang; marked 'Press'	each	0.216	3.35	4.53	1.18	9.06

250 volt grade flush-mounted accessories; including metal back boxes; flexible PVC insulated earth continuity conductor between box and face plate; to backgrounds requiring fixings

	Unit	Hours	Hours £	Mat'ls £	O & P £	Total £
White plastic switch socket outlets						
13 amp; 1 gang	each	0.348	5.39	2.86	1.24	9.49
13 amp; 2 gang	each	0.448	6.94	4.59	1.73	13.26
13 amp; 1 gang including neon indicator	each	0.348	5.39	7.83	1.98	15.21
13 amp; 2 gang including neon indicator	each	0.448	6.94	13.76	3.11	23.81
White plastic switch fused spur connection unit including final connections						
13 amp; double pole	each	0.548	8.49	4.49	1.95	14.93
13 amp; double pole including neon indicator	each	0.548	8.49	10.06	2.78	21.34

	Unit	Hours	Hours £	Mat'ls £	O & P £	Total £
White plastic double pole switches						
20 amp; double pole	each	0.498	7.72	5.78	2.02	15.52
20 amp; double pole including neon indicator	each	0.498	7.72	8.43	2.42	18.57
20 amp; double pole including neon indicator marked 'Water Heater'	each	0.648	10.04	9.83	2.98	22.86
32 amp; double pole including neon indicator	each	0.531	8.23	10.71	2.84	21.78
45 amp; double pole	each	0.531	8.23	9.80	2.70	20.74
45 amp; double pole including neon indicator	each	0.531	8.23	12.36	3.09	23.68
White plastic safety switch socket outlets with RCD						
13 amp; 1 gang; 30 milli am	each	0.431	6.68	38.97	6.85	52.50
13 amp; 2 gang; 30 milli am	each	0.531	8.23	46.18	8.16	62.57
White plastic unswitched socket outlets						
2 amp; 3 pin	each	0.331	5.13	5.75	1.63	12.51
5 amp; 3 pin	each	0.331	5.13	6.60	1.76	13.49
13 amp; 1 gang	each	0.348	5.39	4.24	1.45	11.08
13 amp; 2 gang	each	0.448	6.94	7.18	2.12	16.24
White plastic unswitched fuse spur connection unit including final connector						
13 amp	each	0.548	8.49	6.98	2.32	17.80
13 amp; including neon indicator	each	0.548	8.49	10.17	2.80	21.46

	Unit	Hours	Hours £	Mat'ls £	O & P £	Total £
White plastic cooker control unit; switched						
50 amp; double pole	each	0.614	9.52	10.04	2.93	22.49
50 amp; double pole; marked 'Cooker'	each	0.614	9.52	11.76	3.19	24.47
50 amp; double pole; including neon indicator	each	0.614	9.52	14.58	3.61	27.71
50 amp; double pole; including neon indicator and marked 'Cooker'	each	0.614	9.52	14.07	3.54	27.13
White plastic cooker connection unit including terminals and cable clamp						
connection unit	each	0.364	5.64	5.92	1.73	13.30
White plastic shaver socket outlet plate 240/115 volt						
dual voltage	each	0.514	7.97	30.83	5.82	44.62
250 volt grade flush-mounted accessories; including dry lining back boxes; fixed into aperture						
White plastic switch socket outlets						
13 amp; 1 gang	each	0.283	4.39	2.35	1.01	7.75
13 amp; 2 gang	each	0.333	5.16	4.21	1.41	10.78

	Unit	Hours	Hours £	Mat'ls £	O & P £	Total £
White plastic switch fused spur connection unit including final connections						
13 amp; double pole	each	0.483	7.49	3.98	1.72	13.19
13 amp; double pole including neon indicator	each	0.483	7.49	9.55	2.56	19.59
White plastic double pole switches						
20 amp; double pole	each	0.466	7.22	5.28	1.88	14.38
20 amp; double pole including neon indicator	each	0.466	7.22	7.92	2.27	17.41
20 amp; double pole including neon indicator marked 'Water Heater'	each	0.616	9.55	9.33	2.83	21.71
32 amp; double pole including neon indicator	each	0.499	7.73	9.93	2.65	20.31
45 amp; double pole	each	0.499	7.73	9.02	2.51	19.27
45 amp; double pole including neon indicator	each	0.499	7.73	11.59	2.90	22.22
White plastic safety switch sockets with RCD						
13 amp; 1 gang; 30 milli amp	each	0.316	4.90	38.66	6.53	50.09
13 amp; 2 gang; 30 milli amp	each	0.416	6.45	40.59	7.06	54.09
White plastic unswitched socket outlets						
2 amp; 3 pin	each	0.266	4.12	5.24	1.40	10.77
5 amp; 3 pin	each	0.266	4.12	6.10	1.53	11.76
13 amp; 1 gang	each	0.283	4.39	3.74	1.22	9.35
13 amp; 2 gang	each	0.283	4.39	6.81	1.68	12.88

	Unit	Hours	Hours £	Mat'ls £	O & P £	Total £
White plastic unswitched fuse spur connection unit including final connector						
13 amp	each	0.483	7.49	6.47	2.09	16.05
13 amp; including neon indicator	each	0.483	7.49	10.11	2.64	20.24
White plastic cooker control unit; switched						
50 amp; double pole	each	0.499	7.73	9.45	2.58	19.76
50 amp; double pole; marked 'Cooker'	each	0.499	7.73	9.96	2.65	20.35
50 amp; double pole; including neon indicator	each	0.499	7.73	12.78	3.08	23.59
50 amp; double pole; including neon indicator and marked 'Cooker'	each	0.499	7.73	12.27	3.00	23.01
White plastic cooker connection unit including terminals and cable clamp						
connection unit	each	0.299	4.63	5.15	1.47	11.25
White plastic shaver socket outlet plate 240/115 volt						
dual voltage	each	0.399	6.18	29.03	5.28	40.50

250 volt grade surface-mounted accessories; including moulded plastic back boxes; to backgrounds requiring fixings

	Unit	Hours	Hours £	Mat'ls £	O & P £	Total £
White plastic switch socket outlets						
13 amp; 1 gang	each	0.333	5.16	2.40	1.13	8.70
13 amp; 2 gang	each	0.433	6.71	4.25	1.64	12.61
White plastic switch fused spur connection unit including final connections						
13 amp; double pole	each	0.583	9.04	4.02	1.96	15.01
13 amp; double pole including neon indicator	each	0.583	9.04	9.55	2.79	21.37
White plastic double pole switches						
20 amp; double pole	each	0.466	7.22	5.42	1.90	14.54
20 amp; double pole including neon indicator	each	0.466	7.22	8.05	2.29	17.56
20 amp; double pole including neon indicator marked 'Water Heater'	each	0.616	9.55	9.44	2.85	21.84
32 amp; double pole including neon indicator	each	0.499	7.73	10.05	2.67	20.45
45 amp; double pole	each	0.499	7.73	9.14	2.53	19.41
45 amp; double pole including neon indicator	each	0.499	7.73	11.97	2.96	22.66
White plastic safety switch sockets with RCD						
13 amp; 1 gang; 30 milli amp	each	0.416	6.45	39.41	6.88	52.74
13 amp; 2 gang; 30 milli amp	each	0.466	7.22	46.30	8.03	61.55

	Unit	Hours	Hours £	Mat'ls £	O & P £	Total £
White plastic unswitched socket outlets						
2 amp; 3 pin	each	0.316	4.90	5.39	1.54	11.83
5 amp; 3 pin	each	0.316	4.90	6.24	1.67	12.81
13 amp; 1 gang	each	0.366	5.67	3.89	1.43	11.00
13 amp; 2 gang	each	0.516	8.00	6.94	2.24	17.18
White plastic unswitched fuse spur connection unit including final connector						
13 amp	each	0.483	7.49	6.60	2.11	16.20
13 amp; including neon indicator	each	0.483	7.49	10.23	2.66	20.37
White plastic cooker control unit; switched						
50 amp; double pole	each	0.599	9.28	10.40	2.95	22.64
50 amp; double pole; marked 'Cooker'	each	0.599	9.28	10.91	3.03	23.22
50 amp; double pole; including neon indicator	each	0.599	9.28	13.71	3.45	26.44
50 amp; double pole; including neon indicator and marked 'Cooker'	each	0.599	9.28	13.21	3.37	25.87
White plastic cooker connection unit including terminals and cable clamp						
connection unit	each	0.349	5.41	5.58	1.65	12.64
White plastic shaver socket outlet plate 240/115 volt						
dual voltage	each	0.499	7.73	29.01	5.51	42.26

	Unit	Hours	Hours £	Mat'ls £	O & P £	Total £
White ceiling pull switches including base plate (back box)						
6 amp; 1 way; single pole	each	0.233	3.61	4.21	1.17	8.99
6 amp; 2 way; single pole	each	0.283	4.39	4.79	1.38	10.55
16 amp; 2 way; single pole	each	0.316	4.90	6.55	1.72	13.17
16 amp; 1 way; double pole	each	0.350	5.43	9.41	2.23	17.06
40 amp; 1 way; double pole	each	0.383	5.94	11.78	2.66	20.37
45 amp; 1 way; double pole complete with indicator lanp	each	0.350	5.43	8.26	2.05	15.74
250 volt grade flush-mounted accessories; including cable trunking back boxes; fix into trunking body						
White plastic switch plates; 6 amp rated						
1 gang; 1 way; single pole	each	0.233	3.61	2.60	0.93	7.14
1 gang; 2 way; single pole	each	0.266	4.12	2.77	1.03	7.93
1 gang; 1 way; double pole	each	0.299	4.63	5.90	1.58	12.11
1 gang; marked with 'Bell' symbol	each	0.233	3.61	4.91	1.28	9.80
1 gang; marked 'Press'	each	0.233	3.61	4.91	1.28	9.80
White plastic switch socket outlets						
13 amp; 1 gang	each	0.333	5.16	4.04	1.38	10.58
13 amp; 2 gang	each	0.433	6.71	5.91	1.89	14.51
White plastic switch fused spur connection unit including final connections						
13 amp; double pole	each	0.383	5.94	5.65	1.74	13.32
13 amp; double pole including neon indicator	each	0.383	5.94	11.18	2.57	19.68

	Unit	Hours	Hours £	Mat'ls £	O & P £	Total £
White plastic double pole switches						
20 amp; double pole	each	0.416	6.45	6.93	2.01	15.38
20 amp; double pole including neon indicator	each	0.416	6.45	9.57	2.40	18.42
32 amp; double pole including neon indicator	each	0.449	6.96	11.57	2.78	21.31
45 amp; double pole	each	0.499	7.73	10.66	2.76	21.15
45 amp; double pole including neon indicator	each	0.499	7.73	13.21	3.14	24.09

250 volt grade flush-mounted accessories; including 'Metalclad' range; galvanised back boxes; to backgrounds requiring fixings

	Unit	Hours	Hours £	Mat'ls £	O & P £	Total £
'Metalclad' switch plates; galvanised with white inserts; 6 amp rated						
1 gang; 1 way; single pole	each	0.233	3.61	5.29	1.34	10.24
1 gang; 2 way; single pole	each	0.266	4.12	5.87	1.50	11.49
2 gang; 1 way; single pole	each	0.299	4.63	6.95	1.74	13.32
2 gang; 2 way; single pole	each	0.332	5.15	8.36	2.03	15.53

250 volt grade surface-mounted accessories; including 'Metalclad' range; galvanised back boxes; flexible PVC insulated earth conductor between box and face plate; to backgrounds requiring fixings

	Unit	Hours	Hours £	Mat'ls £	O & P £	Total £
'Metalclad' switch socket outlet plates galvanised with white inserts						
13 amp; 1 gang	each	0.398	6.17	6.93	1.96	15.06
13 amp; 2 gang	each	0.465	7.21	13.32	3.08	23.61

	Unit	Hours	Hours £	Mat'ls £	O & P £	Total £
'Metalclad' switch socket outlet (cont'd)						
13 amp; 1 gang including neon indicator	each	0.394	6.11	9.92	2.40	18.43
13 amp; 2 gang including neon indicator	each	0.465	7.21	18.64	3.88	29.72
'Metalclad' switch fused spur connection unit; galvanised face plate with white inserts including final connections						
13 amp; double pole	each	0.448	6.94	7.51	2.17	16.62
13 amp; double pole including neon indicator	each	0.448	6.94	9.50	2.47	18.91
13 amp; double pole including flex outlet	each	0.598	9.27	8.18	2.62	20.07
13 amp; double pole including neon indicator and flex outlet	each	0.598	9.27	9.88	2.87	22.02
'Metalclad' double pole switches; galvanised face plates with white inserts						
20 amp; including neon indicator	each	0.581	9.01	9.55	2.78	21.34
20 amp; including neon indicator and cord outlet	each	0.714	11.07	9.93	3.15	24.15
50 amp	each	0.464	7.19	8.94	2.42	18.55
50 amp; including neon indicator	each	0.464	7.19	12.14	2.90	22.23
'Metalclad' safety switch sockets; galvanised face plate with white insert and RCD						
13 amp; 1 gang; 30 milli amp	each	0.548	8.49	40.87	7.40	56.77
13 amp; 2 gang; 30 milli amp	each	0.598	9.27	47.39	8.50	65.16

	Unit	Hours	Hours £	Mat'ls £	O & P £	Total £
'Metalclad' unswitched socket outlets; galvanised face plate with white inserts						
13 amp; 1 gang	each	0.398	6.17	5.19	1.70	13.06
13 amp; 2 gang	each	0.465	7.21	9.66	2.53	19.40
'Metalclad' unswitched fused spur connection units; galvanised face plate with white inserts including final connections						
13 amp	each	0.598	9.27	6.86	2.42	18.55
13 amp; including flex outlet	each	0.598	9.27	7.46	2.51	19.24
'Metalclad' switch plates; satin chrome with white inserts; 6 amp rated						
1 gang; 1 way; single pole	each	0.233	3.61	5.06	1.30	9.97
1 gang; 2 way; single pole	each	0.266	4.12	5.12	1.39	10.63
2 gang; 2 way; single pole	each	0.332	5.15	6.83	1.80	13.77
4 gang; 2 way; single pole	each	0.631	9.78	11.67	3.22	24.67
1 gang; 2 way; key switch; single pole including key	each	0.281	4.36	7.97	1.85	14.17
'Metalclad' switch socket outlet plates; satin chrome face plate with white inserts						
13 amp; 1 gang	each	0.398	6.17	6.85	1.95	14.97
13 amp; 2 gang	each	0.465	7.21	12.70	2.99	22.89

	Unit	Hours	Hours £	Mat'ls £	O & P £	Total £
'Metalclad' switch fused spur connection units; satin chrome face plates with white inserts including final connections						
13 amp; double pole	each	0.448	6.94	7.75	2.20	16.90
13 amp; double pole including neon indicator	each	0.448	6.94	9.38	2.45	18.77
13 amp; double pole including flex outlet	each	0.598	9.27	8.40	2.65	20.32
13 amp; double pole including neon indicator and flex outlet	each	0.598	9.27	9.91	2.88	22.06
'Metalclad' double pole switches with satin chrome face plates with white inserts						
30 amp; double pole	each	0.531	8.23	7.44	2.35	18.02
30 amp; double pole including neon indicator	each	0.531	8.23	9.28	2.63	20.14
30 amp; double pole including flex outlet	each	0.681	10.56	6.60	2.57	19.73
45 amp; double pole	each	0.531	8.23	10.37	2.79	21.39
45 amp; double pole including neon indicator	each	0.531	8.23	11.80	3.00	23.04
250 volt grade surface-mounted weather proof accessories including polycarbonate back box; to backgrounds requiring fixings						
White polycarbonate switch plate to IP56; 10 amp rated						
1 gang; 1 way; single pole	each	0.283	4.39	11.09	2.32	17.80
1 gang; 2 way; single pole	each	0.316	4.90	12.49	2.61	20.00
1 gang; 2 way; single pole marked with 'Bell' symbol	each	0.316	4.90	15.74	3.10	23.73

	Unit	Hours	Hours £	Mat'ls £	O & P £	Total £
1 gang; 2 way; single pole marked 'Press'	each	0.316	4.90	15.74	3.10	23.73
White polycarbonate socket outlets with 30 milli amp RCD						
13 amp; 1 gang	each	0.600	9.30	80.78	13.51	103.59
White polycarbonate unswitched socket outlets						
13 amp; 1 gang	each	0.550	8.53	19.32	4.18	32.02
13 amp; 2 gang	each	0.700	10.85	33.44	6.64	50.93
Sound and vision flush-mounted accessories; including metal back boxes; to backgrounds requiring fixings						
White plastic TV aerial (coaxial) socket outlet plates						
1-way direct connection	each	0.399	6.18	5.92	1.82	13.92
2-way direct connection	each	0.482	7.47	7.88	2.30	17.65
1-way isolated UHF/VHF	each	0.399	6.18	9.76	2.39	18.34
2-way isolated UHF/VHF	each	0.482	7.47	13.05	3.08	23.60
Sound and vision flush-mounted accessories; including dry lining boxes; fixed into apertures						
White plastic telephone socket outlet plates						
single master outlet	each	0.230	3.57	7.43	1.65	12.64
single secondary outlet	each	0.230	3.57	5.40	1.34	10.31

	Unit	Hours	Hours £	Mat'ls £	O & P £	Total £
Sound and vision flush-mounted accessories; including dry lining boxes; to backgrounds requiring fixings						
White plastic TV aerial (coaxial) socket outlet plates						
1-way direct connection	each	0.366	5.67	5.88	1.73	13.29
2-way direct connection	each	0.449	6.96	7.83	2.22	17.01
1-way isolated UHF/VHF	each	0.366	5.67	9.71	2.31	17.69
2-way isolated UHF/VHF	each	0.449	6.96	13.01	3.00	22.96
White plastic telephone socket outlet plates						
single master outlet	each	0.197	3.05	7.42	1.57	12.04
single secondary outlet	each	0.197	3.05	5.39	1.27	9.71
Sound and vision flush-mounted accessories; including cable trunking back boxes; fixed into trunking body						
White plastic TV aerial (coaxial) socket outlet plates						
1-way direct connection	each	0.416	6.45	7.07	2.03	15.55
2-way direct connection	each	0.499	7.73	9.02	2.51	19.27
1-way isolated UHF/VHF	each	0.416	6.45	10.90	2.60	19.95
2-way isolated UHF/VHF	each	0.499	7.73	14.19	3.29	25.21
White plastic telephone socket outlet plates						
single master outlet	each	0.247	3.83	8.61	1.87	14.30
single secondary outlet	each	0.247	3.83	6.59	1.56	11.98

	Unit	Hours	Hours £	Mat'ls £	O & P £	Total £
Time clock and programmers						
240 volt grade surface-mounted domestic central heating programmers; as supplied by 'Newlec'						
single circuit; four programme controller; to backgrounds requiring fixings	each	0.500	7.75	36.50	6.64	50.89
sixteen programme controller; to backgrounds requiring fixings	each	0.500	7.75	43.40	7.67	58.82
single channel, 7 day programmer including battery back up; to backgrounds requiring fixings	each	0.500	7.75	35.91	6.55	50.21
two channel, 24 hour programmer; to backgrounds requiring fixings	each	0.500	7.75	36.50	6.64	50.89
two channel, 7 day programmer; to backgrounds requiring fixings	each	0.500	7.75	39.39	7.07	54.21
240 volt grade surface-mounted domestic water heating programmers; as supplied by 'Newlec'						
'Economy 7' quartz programmer; to backgrounds requiring fixings	each	0.583	9.04	60.83	10.48	80.35
'Economy' electronic programmer including battery back-up; to backgrounds requiring fixings	each	0.583	9.04	61.71	10.61	81.36

	Unit	Hours	Hours £	Mat'ls £	O & P £	Total £
Y41 FANS						
Bathroom and toilet extract fans as supplied by 'Newlec'						
Axial fan; double insulated; IP44 rated; fixing into aperture aperture measured elsewhere and including 2 core 0.75mm2 heat resistant flexible cable not exceeding 1000mm long and final connections						
100mm diameter	each	1.250	19.38	16.04	5.31	40.73
100mm diameter complete with timer (3-20minutes)	each	1.250	19.38	22.29	6.25	47.91
100mm diameter complete with shutters and adjustable timer (3-20 minutes)	each	1.250	19.38	25.92	6.79	52.09
100mm diameter kit comprising fan complete with timer (3-20, flexible minutes), duct (3m), external wall grille and duct ties	each	1.250	19.38	24.37	6.56	50.31
100mm diameter complete with humidistat and pull switch	each	1.250	19.38	30.58	7.49	57.45
100mm diameter complete with shutters and condensation control	each	1.250	19.38	41.87	9.19	70.43
100mm diameter complete with shutters and pull cord	each	1.250	19.38	21.78	6.17	47.33
100mm diameter inline shower kit comprising fan, 3 metres ducting, inlet grille, soffit grille and duct ties	each	1.500	23.25	24.84	7.21	55.30
Ventilation accessories, fix to axial fan/backgrounds requiring fixings						
100mm diameter flexible ducting 3 metres length	each	0.250	3.88	10.20	2.11	16.19

	Unit	Hours	Hours £	Mat'ls £	O & P £	Total £
100mm diameter exterior wall grille	each	0.250	3.88	4.05	1.19	9.11
100mm diameter wall vent kit	each	0.250	3.88	7.12	1.65	12.64
100mm diameter window kit	each	0.250	3.88	7.43	1.70	13.00
hose securing ties	each	0.250	3.88	2.15	0.90	6.93

Sundry Accessories - Fan Controller

250 volt grade flush-mounted accessories; including metal back boxes; flexible PVC insulated earth continuity conductor box between box and face plate; to backgrounds requiring fixings

White plastic fan isolater switch

	Unit	Hours	Hours £	Mat'ls £	O & P £	Total £
6 amp; triple pole isolater switch engraved 'Fan'	each	0.833	12.91	5.99	2.84	21.74

250 volt grade flush-mounted accessories; including dry lining back boxes; fitted into aperture

White plastic fan isolater switch

	Unit	Hours	Hours £	Mat'ls £	O & P £	Total £
6 amp; triple pole isolater switch engraved 'Fan'	each	0.815	12.63	5.95	2.79	21.37

250 volt grade flush-mounted accessories; including moulded plastic back boxes; to backgrounds requiring fixings

White plastic fan isolater switch

	Unit	Hours	Hours £	Mat'ls £	O & P £	Total £
6 amp; triple pole isolater switch engraved 'Fan'	each	0.850	13.18	6.04	2.88	22.10

	Unit	Hours	Hours £	Mat'ls £	O & P £	Total £
Y80 EARTHING AND BONDING						
High conductivity copper tape to BS 1432; bare						
To backgrounds requiring fixings; fixings measured elsewhere						
25 × 3mm; including dressing	m	0.141	2.19	4.99	1.08	8.25
Laid in trench						
25 × 3mm	m	0.100	1.55	4.99	0.98	7.52
High conductivity copper tape to BS 5225; PVC sheathed						
To backgrounds requiring fixings; fixings measured elsewhere						
25 × 3mm; including dressing	m	0.141	2.19	6.50	1.30	9.99
Laid in trench						
25 × 3mm	m	0.100	1.55	6.50	1.21	9.26
Copper tape ancillaries						
Non-metallic DC clips; high quality polypropylene; to backgrounds requiring fixings						
25 × 3mm	each	0.041	0.64	0.52	0.17	1.33
Metallic DC clips; high grade copper; to backgrounds requiring fixings						
25 × 3mm	each	0.050	0.78	1.39	0.32	2.49

	Unit	Hours	Hours £	Mat'ls £	O & P £	Total £
Square gunmetal clamps; to backgrounds requiring fixings						
25 × 3mm	each	0.166	2.57	3.38	0.89	6.85
Copperbond earth rod; complete with coupling and driving stud						
1200mm in length; driven into average ground						
16mm diameter	each	0.250	3.88	8.58	1.87	14.32
Earth rod ancillaries						
Lightweight earth pit with lockable jam-free lid						
150 × 150 × 200mm deep; excavation	each	0.500	7.75	22.25	4.50	34.50
Silicon aluminium bronze body with phosphor bronze screw						
16mm; rod to cable clamp	each	0.166	2.57	1.63	0.63	4.83
16mm; rod to tape clamp	each	0.166	2.57	3.20	0.87	6.64
Earth cable ancillaries						
Stainless steel earth clamp; BS 951; complete with locknut and warning label; pipe size 12 to 32mm diameter						
indoor; forming around pipe; 2.5mm2 to 10mm2 terminal	each	0.116	1.80	0.68	0.37	2.85
outdoor; forming around pipe; 2.5mm2 to 10mm2 terminal	each	0.116	1.80	0.75	0.38	2.93
heavy duty outdoor; forming around pipe; 2.5mm2 to 16mm2 terminal	each	0.125	1.94	1.10	0.46	3.49

	Unit	Hours	Hours £	Mat'ls £	O & P £	Total £
Stainless steel earth clamp suitable bonding radiators and the like complete with clamp and safety label; supplied in packs of 10						
indoor; clean up and fix; 1.5mm2 to 4mm2 crimp lug	each	0.100	1.55	8.75	1.55	11.85

	Unit	Hours	Hours £	Mat'ls £	O & P £	Total £
V51 LOCAL ELECTRIC HEATING UNITS						
Convector heaters; as manufactured by 'Creda' and of their 'TPR' range						
Panel heaters; including final connection to the power supply; to backgrounds requiring fixings						
600 watt; 420 × 444 × 93mm	each	0.916	14.20	44.96	8.87	68.03
1000 watt; 420 × 614 × 93mm	each	1.083	16.79	48.93	9.86	75.57
1250 watt; 420 × 614 × 93mm	each	1.083	16.79	54.67	10.72	82.17
1500 watt; 420 × 682 × 93mm	each	1.083	16.79	59.80	11.49	88.07
2000 watt; 420 × 852 × 93mm	each	1.116	17.30	63.76	12.16	93.22
Panel heaters complete with timer and Quartz Electronic programmer; including final connections to power supply; to backgrounds requiring fixings						
600 watt; 420 × 444 × 93mm	each	0.916	14.20	61.86	11.41	87.47
1000 watt; 420 × 614 × 93mm	each	1.083	16.79	66.08	12.43	95.30
1250 watt; 420 × 614 × 93mm	each	1.083	16.79	72.02	13.32	102.13
1500 watt; 420 × 682 × 93mm	each	1.083	16.79	77.37	14.12	108.28
2000 watt; 420 × 852 × 93mm	each	1.116	17.30	82.18	14.92	114.40
Tubular heaters as supplied by 'Newlec' and complete with wall bracket						
Tubular heaters complete with auto reset cut-outs including final connections to the power supply; to backgrounds requiring fixings						

	Unit	Hours	Hours £	Mat'ls £	O & P £	Total £
Tubular heaters (cont'd)						
2 feet; 120watt	each	1.000	15.50	14.60	4.52	34.62
4 feet; 240watt	each	1.000	15.50	19.77	5.29	40.56
6 feet; 360watt	each	1.250	19.38	24.95	6.65	50.97
Tubular heater wire guards, galvanised finish; to backgrounds requiring fixings						
2 feet; 29" × 6" × 6"	each	0.332	5.15	23.45	4.29	32.89
4 feet; 53" × 6" × 6"	each	0.332	5.15	35.87	6.15	47.17
6 feet; 74" × 6" × 6"	each	0.332	5.15	46.22	7.70	59.07
Storage heaters as manufactured by 'Creda' and of their 'TSR Sensor Plus' range						
Storage heaters complete with automatic control including final connections to the power supply; to backgrounds requiring fixings						
900 watt; 725 × 335 × 167mm	each	1.000	15.50	114.89	19.56	149.95
1700 watt; 725 × 788 × 167mm	each	1.000	15.50	158.36	26.08	199.94
2500 watt; 725 × 788 × 167mm	each	1.250	19.38	197.69	32.56	249.62
3400 watt; 725 × 1016 × 167mm	each	1.500	23.25	242.19	39.82	305.26
Storage heaters as manufactured by 'Creda' and of their 'TSR Supaslim Combi' range						
Storage heaters complete with fan system, thermostatic control two heat options including final connections to the power supply; to backgrounds requiring fixings						

	Unit	Hours	Hours £	Mat'ls £	O & P £	Total £
Storage heaters (cont'd)						
1700 watt; 744 × 560 × 167mm	each	1.000	15.50	198.72	32.13	246.35
2500 watt; 744 × 788 × 167mm	each	1.250	19.38	347.37	55.01	421.76
3400 watt; 744 × 1016 × 167mm	each	1.500	23.25	291.87	47.27	362.39

Note: Add extra time for installation of storage heaters above first floor level

Instant water boilers as manufactured by 'Zip' Hydroboil Series 2000

White stove enamel; twin chamber system complete with electro-mechanical controls, power indicator, safety cut-out and tap; to backgrounds requiring fixings including final connection to cold water supply

	Unit	Hours	Hours £	Mat'ls £	O & P £	Total £
1.5 kW rating; 2 litre capacity (9 cups)	each	1.916	29.70	369.20	59.83	458.73
1.5 kW rating; 3 litre capacity (18 cups)	each	1.916	29.70	405.60	65.29	500.59
2.4 kW rating; 5 litre capacity (30 cups)	each	1.916	29.70	462.80	73.87	566.37
2.4 kW rating; 7 litre capacity (45 cups)	each	1.916	29.70	514.80	81.67	626.17
3.0 kW rating; 10 litre capacity (60 cups)	each	1.916	29.70	623.41	97.97	751.07
3.0 kW rating; 15 litre capacity (90 cups)	each	1.916	29.70	741.94	115.75	887.38
3.0 kW rating; 25 litre capacity (150 cups)	each	1.916	29.70	900.28	139.50	1069.47

	Unit	Hours	Hours £	Mat'ls £	O & P £	Total £
Stainless steel; twin chamber system complete with electro-mechanical controls, power indicator, safety cut-out and tap; to backgrounds requiring fixings including final connection to cold water supply						
2.4 kW rating; 5 litre capacity (30 cups)	each	1.916	29.70	514.34	81.61	625.64
2.4 kW rating; 7 litre capacity (45 cups)	each	1.916	29.70	576.77	90.97	697.44
3.0 kW rating; 10 litre capacity (60 cups)	each	1.916	29.70	721.54	112.69	863.92
3.0 kW rating; 15 litre capacity (90 cups)	each	1.916	29.70	866.31	134.40	1030.41
3.0 kW rating; 25 litre capacity (150 cups)	each	1.916	29.70	1013.79	156.52	1200.01
6.0 kW rating; 50 litre capacity (300 cups)	each	1.916	29.70	1335.00	204.70	1569.40

Immersion heaters

	Unit	Hours	Hours £	Mat'ls £	O & P £	Total £
Domestic immersion heaters and fitted to existing standard boss including 3 core 2.5mm2 heat resistant flexible cable not exceeding 1000mm long						
3 kW; 11 in long including thermostat	each	1.032	16.00	14.64	4.60	35.23
3 kW; 27 in long including thermostat	each	1.032	16.00	15.44	4.72	36.15

	Unit	Hours	Hours £	Mat'ls £	O & P £	Total £
Industrial immersion heaters single phase and fitted to existing standard boss including 3 core 2.5mm2 heat resistant flexible cable not exceeding 1000mm long						
2 kW; 11 in long	each	1.032	16.00	84.99	15.15	116.13
3 kW; 11 in long	each	1.032	16.00	109.21	18.78	143.99
3 kW; 30 in long	each	1.032	16.00	113.19	19.38	148.56
4 kW; 11 in long	each	1.032	16.00	111.48	19.12	146.60
4 kW; 16 in long	each	1.032	16.00	112.71	19.31	148.01
Electric showers as manufact-ured by Newlec						
Electric showers with temperature control and power settings complete with flexible hose, riser rail, shower hand set and electrical connection only (excludes water connection)						
7.2 kW; shower unit	each	2.066	32.02	51.70	12.56	96.28
8.5 kW; shower unit	each	2.066	32.02	56.93	13.34	102.30
8.5 kW; 'Calypso Plus' shower unit	each	2.066	32.02	89.01	18.15	139.19
9.5 kW; 'Calypso Plus' shower unit	each	2.066	32.02	92.12	18.62	142.76

	Unit	Hours	Hours £	Mat'ls £	O & P £	Total £
W30 DATA TRANSMISSION						
Communication non-joiner cables						
PVC sheath, tinned annealed copper conductors, complying with BT specification CW 1308						
to backgrounds requiring fixings						
4 wire 1.5; 2 pr	m	0.033	0.51	0.09	0.09	0.69
6 wire 1.5; 3 pr	m	0.041	0.64	0.12	0.11	0.87
8 wire 1.5; 4 pr	m	0.046	0.71	0.14	0.13	0.98
12 wire 1.5; 6 pr	m	0.050	0.78	0.28	0.16	1.21
laid into trunking						
4 wire 1.5; 2 pr	m	0.016	0.25	0.09	0.05	0.39
6 wire 1.5; 3 pr	m	0.025	0.39	0.12	0.08	0.58
8 wire 1.5; 4 pr	m	0.025	0.39	0.14	0.08	0.61
12 wire 1.5; 6 pr	m	0.038	0.59	0.28	0.13	1.00
Co-axiel cable; annealed copper conductor with cellular polythene insulation, overall PVC sheath, braided copper screen, 750ohm impedance; semi air spaced low lose TV downlead; to backgrounds requiring fixings						
TV downlead 1mm2	m	0.033	0.51	0.19	0.11	0.81
Complete termination comprising aluminium co-axial plug and soldering connection						
co-axial plug	each	0.100	1.55	0.27	0.27	2.09

	Unit	Hours	Hours £	Mat'ls £	O & P £	Total £
Satellite cable; annealed copper conductor with cellular polythene insulation, overall PVC sheath, 64 stranded copper and braided screen, 75 ohm impedance; impedance; semi air spaced low lose satellite downlead; to backgrounds requiring fixings						
satellite downlead 1mm2	m	0.033	0.51	0.20	0.11	0.82
Complete termination comprising satellite 'F' plug and crimped connection						
co-axial plug	each	0.100	1.55	0.49	0.31	2.35
Cable sundries						
Stainless steel/brass earth clamp, BS 951, complete with lock nuts and warning label; pipe size 12 to 32mm diameter						
indoor, forming around pipe, 2.5mm2 to 10mm2 terminal	each	0.233	3.61	0.59	0.63	4.83
outdoor, forming around pipe, 2.5mm2 to 16mm2 terminal	each	0.233	3.61	0.89	0.68	5.18
Cable termination and connection; cutting and dressing cable, fit gland and connect to equipment, supply and fit identity markers and connect to equipment including the removal and reinstallation of gland plate						

	Unit	Hours	Hours £	Mat'ls £	O & P £	Total £
XLPE/SWA/LSF cable 0.8mm to 1.13mm diameter						
2 core cable	m	1.000	15.50	0.12	2.34	17.96
3 core cable	m	1.111	17.22	0.18	2.61	20.01
4 core cable	m	1.222	18.94	0.24	2.88	22.06
Unarmoured PVC sheathed cable 0.8mm diameter						
2 core cable	m	0.666	10.32	0.12	1.57	12.01
3 core cable	m	0.722	11.19	0.18	1.71	13.08
4 core cable	m	0.777	12.04	0.24	1.84	14.13

	Unit	Hours	Hours £	Mat'ls £	O & P £	Total £
W41 SECURITY						
INTRUDER ALARM SYSTEMS						
Wired intruder alarm equipment as supplied by Newlec						
Surface-mounted intruder alarm systems meeting regulations ETSI 300/220 and complete with detector batteries; to backgrounds requiring fixings						
8 zone programmable control panel with 12 volt rechargable battery pack, keypad with PA button combination	each	2.000	31.00	38.40	10.41	79.81
Bell box complete with strobe, anti-tamper protection and integral battery	each	1.066	16.52	31.21	7.16	54.89
Dummy bell box	each	0.500	7.75	6.32	2.11	16.18
PTR detector unit complete with selectable pulse unit	each	0.666	10.32	12.24	3.38	25.95
Magnetic door contacts	each	1.166	18.07	1.32	2.91	22.30
Remote keypad, backlit and complete with PA buttons	each	1.000	15.50	17.16	4.90	37.56
Low voltage alarm cable, 6 core PVC sheathed to comply with BS 4737	m	0.033	0.51	0.17	0.10	0.78

	Unit	Hours	Hours £	Mat'ls £	O & P £	Total £
Surface-mounted 1/3"CCD fixed dome camera complete with mini-dome (4.5" diameter), 4mm lens, electronic iris and direct DIN connection; to backgrounds requiring fixings						
colour dome camera	each	0.750	11.63	160.15	25.77	197.54
monochrome dome camera	each	0.750	11.63	95.15	16.02	122.79
External weather-proof camera housing complete with adjustable high-strength bracket; to backgrounds requiring fixings						
external housing	each	0.666	10.32	74.15	12.67	97.14
Closed-circuit television equipment as supplied by Newlec						
Surface-mounted 1/3"CCD camera complete with built in microphone and speaker, 4mm lens camera and 20m of cable (internal use), external housing measured elsewhere; to backgrounds requiring fixings						
colour camera	each	1.083	16.79	169.15	27.89	213.83
monochrome camera	each	1.083	16.79	105.15	18.29	140.23

	Unit	Hours	Hours £	Mat'ls £	O & P £	Total £
W50 FIRE DETECTION AND ALARM						
Fire alarm equipment as supplied by Newlec						
Surface mounted fire alarm panel to BS 5839 Parts 1 and 4 and EN 54 Parts 2 and 4 complete with standby batteries; to backgrounds requiring fixings						
1 zone fire alarm panel	each	1.500	23.25	67.69	13.64	104.58
2 zone fire alarm panel	each	2.000	31.00	88.39	17.91	137.30
4 zone fire alarm panel	each	2.000	31.00	140.40	25.71	197.11
Surface-mounted detectors including base; fixed to conduit box						
optical smoke detector	each	0.330	5.12	24.41	4.43	33.95
ionisation smoke detector	each	0.330	5.12	24.41	4.43	33.95
high temperature fixed heat detector	each	0.330	5.12	12.23	2.60	19.95
rate of rise heat detector	each	0.660	10.23	12.23	3.37	25.83
Surface-mounted sundries; fixed to conduit box measured elsewhere						
diode detector base	each	0.330	5.12	4.42	1.43	10.97
Surface mounted call point including back box; to backgrounds requiring fixings						
break glass call point	each	0.500	7.75	7.56	2.30	17.61

	Unit	Hours	Hours £	Mat'ls £	O & P £	Total £
Flush mounted call point; to backgrounds requiring fixings						
break glass call point	each	0.500	7.75	7.26	2.25	17.26
Call point sundries; place in position						
replacement spare glasses supplied in packs of 5	each	0.250	3.88	6.04	1.49	11.40
keys; supplied in packs of 10	each	0.025	0.39	7.05	1.12	8.55
Surface-mounted sounders including back box; to backgrounds requiring fixings						
6 in diameter 24 volt DC bel	each	0.750	11.63	14.67	3.94	30.24
24 volt low profile sounder	each	0.750	11.63	15.18	4.02	30.83
240/9 volt grade surface-mounted detector including pattress and battery hard wired into the lighting circuit; to backgrounds requiring fixings						
ionisation smoke alarm	each	0.500	7.75	14.44	3.33	25.52
Surface-mounted single loop analogue addressable fire alarm panel to BS 5839 Parts 1 and 4 complete with standby batteries; to backgrounds requiring fixings						
16 zone fire alarm panel	each	2.000	31.00	517.50	82.28	630.78
(Add for each zone above 4 zones)						

	Unit	Hours	Hours £	Mat'ls £	O & P £	Total £
Surface-mounted single loop repeater panel; to backgrounds requiring fixings						
repeater panel	each	7.000	108.50	517.50	93.90	719.90
Surface-mounted analogue addressable detectors including base; fixed to conduit box						
photo electric smoke detector	each	0.330	5.12	33.50	5.79	44.41
ionisation smoke detector	each	0.330	5.12	33.50	5.79	44.41
heat detector	each	0.330	5.12	31.47	5.49	42.07
Surface-mounted analogue addressable sundries; fixed to conduit box measured elsewhere						
common detector base	each	0.330	5.12	3.55	1.30	9.96
Surface-mounted analogue addressable call point including back box; to backgrounds requiring fixings						
manual call point	each	0.500	7.75	32.48	6.03	46.26
manual call point IP65	each	0.666	10.32	87.25	14.64	112.21
Manual call point sundries; place in position						
replacement spare glasses supplied in packs of 5	each	0.250	3.88	6.04	1.49	11.40

	Unit	Hours	Hours £	Mat'ls £	O & P £	Total £
Surface-mounted analogue addressable sounders including back box; to backgrounds requiring fixings						
low current sounder	each	0.750	11.63	53.80	9.81	75.24
(Commissioning of this system has been excluded and should be carried out by a Newlec specialist)						
Carbon monoxide alarm; as manufactured by 'AICO'						
250 volt grade surface-mounted detector including pattress; hard wired into power circuit via fuse connection unit, measured elsewhere; to backgrounds requiring fixings						
carbon monoxide alarm, mains wired	each	0.666	10.32	36.87	7.08	54.27
carbon monoxide alarm, mains wired, interlinkable	each	0.666	10.32	44.87	8.28	63.47

	Unit	Hours	Hours £	Mat'ls £	O & P £	Total £
BUILDER'S WORK IN CONNECTION WITH ELECTRICAL INSTALLATION						
The following rates are applicable where there is no general contractor on site and the builder's work and making good is carried out by the electrical contractor						
Wall chases						
Mark out chases on wall and cut by hand						
brickwork; 25mm deep × 15 to 20mm wide	m	0.917	14.21	0.00	2.13	16.35
brickwork; 25mm deep × 25 to 50mm wide	m	1.117	17.31	0.00	2.60	19.91
blockwork; 25mm deep × 15 to 20mm wide	m	0.600	9.30	0.00	1.40	10.70
blockwork; 25mm deep × 25 to 50mm wide	m	0.900	13.95	0.00	2.09	16.04
Accessory apertures						
Mark out apertures for boxes and cut by hand						
brickwork; 15mm deep × 75mm square	each	0.580	8.99	0.00	1.35	10.34
brickwork; 25mm deep × 75mm square	each	0.875	13.56	0.00	2.03	15.60
brickwork; 25mm deep × 135mm long × 75mm wide	each	1.313	20.35	0.00	3.05	23.40

	Unit	Hours	Hours £	Mat'ls £	O & P £	Total £
brickwork; 40mm deep × 75mm square	each	1.455	22.55	0.00	3.38	25.94
blockwork; 15mm deep × 75mm square	each	0.350	5.43	0.00	0.81	6.24
blockwork; 25mm deep × 75mm square	each	0.517	8.01	0.00	1.20	9.22
blockwork; 25mm deep × 135mm long × 75mm wide	each	0.858	13.30	0.00	1.99	15.29
brickwork; 40mm deep × 75mm square	each	0.850	13.18	0.00	1.98	15.15
plasterboard; 12mm deep × 75mm square	each	0.250	3.88	0.00	0.58	4.46
plasterboard; 12mm deep × 135mm long × 75mm wide	each	0.283	4.39	0.00	0.66	5.04

Holes in woodwork

Identify location on joist, mark out and form hole using electric power drill

	Unit	Hours	Hours £	Mat'ls £	O & P £	Total £
38mm thickness of joist						
12.5mm diameter	each	0.020	0.31	0.00	0.05	0.36
20mm diameter	each	0.020	0.31	0.00	0.05	0.36
25mm diameter	each	0.020	0.31	0.00	0.05	0.36
50mm thickness of joist						
12.5mm diameter	each	0.029	0.45	0.00	0.07	0.52
20mm diameter	each	0.029	0.45	0.00	0.07	0.52
25mm diameter	each	0.029	0.45	0.00	0.07	0.52
	each	0.029	0.45	0.00	0.07	0.52

	Unit	Hours	Hours £	Mat'ls £	O & P £	Total £
75mm thickness of joist						
12.5mm diameter	each	0.038	0.59	0.00	0.09	0.68
20mm diameter	each	0.038	0.59	0.00	0.09	0.68
25mm diameter	each	0.038	0.59	0.00	0.09	0.68
38mm diameter	each	0.038	0.59	0.00	0.09	0.68
Lifting flooring; wood						
Lifting softwood tongued and grooved flooring using hand tools to cut tongue and lifting of boards; (saw cut across board measured elsewhere)	m	0.117	1.81	0.00	0.27	2.09
Lifting softwood butt-jointed flooring using hand tools to raise flooring (saw cut across board measured elsewhere)	m	0.094	1.46	0.00	0.22	1.68
extra over; saw cut (2 nr) across flooring not exeeding 150mm wide	each	0.233	3.61	0.00	0.54	4.15
extra over; forming trap in timber flooring comprising two cuts across flooring not exceeding 150mm wide and fixing with 4 nr screws with countersunk heads	each	0.433	6.71	0.00	1.01	7.72
Relay softwood flooring (tongued and grooved and/or butt jointed); fixing with wood screws with countersunk heads	m	0.100	1.55	0.00	0.23	1.78
Trench work						
Excavating soil by hand assuming that the land has been cultivated and is of medium quality with minor rubble	m3	2.666	41.32	0.00	6.20	47.52

	Unit	Hours	Hours £	Mat'ls £	O & P £	Total £
Extra over for heavy soil of building site standard	m3	1.084	16.80	0.00	2.52	19.32
Back fill of excavation with selected material on bottom of trench	m3	0.660	10.23	0.00	1.53	11.76
Warning tiles, markers and tape						
Tiles and markers						
Off loading and laying precast concrete warning tiles into trenches; average size of tile 914 × 229 × 63mm	m	0.125	1.94	0.00	0.29	2.23
Off loading and placing precast concrete marker blocks into excavation; average size of block 305 × 305 × 152mm	each	0.116	1.80	0.00	0.27	2.07
Tape						
Laying of 150mm wide PVC marker tape into trench	m	0.033	0.51	0.00	0.08	0.59

	Unit	Hours	Hours £	Mat'ls £	O & P £	Total £
SUNDRY WORK						
Door bell system						
Door chimes and bell push including associated wiring and protection; to backgrounds requiring fixings						
Door chimes	each	0.500	7.75	8.99	2.51	19.25
External push	each	0.250	3.88	2.10	0.90	6.87
Bell cable (10m)	each	0.330	5.12	1.50	0.99	7.61
PVC capping (2m)	each	0.132	2.05	0.76	0.42	3.23

Part Two

PROJECT COSTS

Upgrading of electrical services

Rewiring

New electrical installation

UPGRADING OF ELECTRICAL SERVICES

The work consists of testing and inspecting meter tails, main earth conductors, bonding conductors and circuit protective conductors and final sub-circuits, and establish their size and designation. The test is to indicate whether the cables are fit for purpose and comply with the IEE Wiring Regulations.

The visual inspection is to comply with the recommendations outlined in the IEE Wiring Regulations, BS 7671, and covers consumer units, meter tails, earths final sub-circuits, electrical accessories and the like.

The Contractor may not switch 'off' supplies to the properties for long periods without giving notice and an assessment of how long the power will be off.

It is the Contractor's responsibility to restore/reset items of equipment and appliances that have been affected by the direct result of switching 'off' the power. Each existing final circuit is to be inspected visually including all equipment associated with the circuits, i.e. luminaires, switches, socket outlets, cooker outlets and the like.

The types and quantities of accessories are to be scheduled, per property, and will be accompanied by a written report on the findings and recommendations. The works are carried out whilst the dwellings are occupied. A new split load consumer unit is installed in each dwelling complete with an integral isolator, RCCB, MCBs and blanks, and all circuits shall be clearly identified and scheduled within the consumer unit and comprise the following:

100 Amp DP disconnection switch	1 nr
80 Amp; 30 milli amp split load RCCB	1 nr
6 Amp MCB (lighting)	1 nr
16 Amp MCB (boiler or immersion heater)	1 nr
32 Amp MCB (ring main)	1 nr
45 Amp MCB (cooker)	1 nr
Blank plates	various

The meter tails, main earth, main equipotential earth bonding and cross-bonding of water services are upgraded, including new earth clips. All the electrical accessories, i.e. switch plates, socket outlet plates, fuse connection plates, immersion switch plates, ceiling roses etc, are all replaced regardless of condition including the introduction of a zones 1/3 luminaire in the bathroom.

If it is found necesary, due to the introduction of additional lighting, the switch plate is to be upgraded, i.e. 2 gang to 3 gang. The tenant is given the choice of a fluorescent luminaire with perspex diffuser or a ceiling pendant in the kitchen.

Each property is furnished with smoke detectors complete with rechargeable batteries, hard wired to the local lighting circuit and ceiling-mounted carbon monoxide detectors are also supplied to those properties with gas fires. These are complete with rechargeable batteries, hard wired to the local power circuit via an unswitched fuse connection unit.

Rates, particularly cables, reflect one-off property alterations, so that smaller discounts would be obtained, particularly on cables, i.e. 50% to 60%.
Lighting in the bathroom shall comply with IEE Regulations, i.e., zoning. For the purposes of the projects in this publication, the assumption has been made that all luminaires fall within zones 1 to 3, and therefore the fitting selected is totally-enclosed. Some tenants may want their own centre and and wall lights re-installed and allowances for disconnecting, testing, where necessary, and re-installation are included.

On completion the Contractor shall ensure that each property is furnished with the following general power requirements, as a minimum:

Area	13 amp switch socket outlet	13 amp fuse connection unit with switch and indicator	13 amp fuse connection unit with switch
Hall	1 single		
Lounge	3 twin		
Dining room	2 twin		
Kitchen	4 twin		1 (fan)
	3 single	3	
Landing	1 single		
Bedrooms	1 twin per bed space		

The proposed installation should be discussed with the tenants and their wishes complied with, e.g. additional sockets over those proposed or a reduction of sockets in areas that have recently been decorated. All cables are installed flush and are protected by suitably sized heavy-duty plastic capping and/or oval conduit complete with fixings. No re-wiring is to be carried out unless tests indicate that this is necessary or if additional outlets are required, i.e. lighting, power, etc.

The low-level socket outlets within the kitchen are mounted 200mm AFFL and controlled via a 13 amp fused connection unit double-pole indicator switch, mounted above the the worktop at the same level as the existing socket outlets to give a uniform appearance. The low-level cooker outlet plate is mounted at the rear of the cooker and at 200mm AFFL. The Contractor will include for re-connecting the cooker to the new low-level outlet plate.

Following the final testing of the installation the Contractor shall demonstrate to each tenant the operation of the RCCB and MCBs and leave a leaflet in each propertry covering this operation for future use.

	Quantity	Unit	Rate	Total £
ONE BEDROOM, ONE PERSON FLAT				
TESTING				
Electric tests				
100% of circuits		Item		18.00
extra for visual inspection		Item		12.00
extra for complete schedules and report		Item		12.00
REMEDIAL WORK				
LV/HV CABLES AND WIRING				
600/1000 volt grade PVC insulated, PVC sheathed cable; single core				
Remove existing meter tails and renew including liaison with local electricity authority/shipper for accessing meter and re-sealing on completion				
25mm2		Item		27.00
600/1000 volt grade PVC insulated, PVC sheathed; including PVC capping where necessary				
Drawn into voids or chases or clipped to backgrounds; twin and earth cabling				
1.5mm2	10	m	2.28	22.80
2.5mm2	20	m	2.74	54.80
6.0mm2	15	m	7.43	111.45
		Carried forward		258.05

	Quantity	Unit	Rate	Total £
		Brought forward		258.05
Drawn into voids or chases or clipped to backgrounds; three core and earth cabling				
1.5mm2	5	m	4.42	22.10
300/500 volt grade, PVC insulated PVC sheathed heat resistant circular cables; copper stranded conductors; BS 6141; in tails (not exceeding 1 metre long) including termination at both ends				
Remove existing immersion heater cable (in tails) and renew				
3 core; 2.5mm2		Item		29.60
EARTHING AND BONDING COMPONENTS				
Copper earth connection including copper insulated cables and final connection and PVC capping where necessary				
Remove main earth between incoming main and consumer unit and renew				
16mm2		Item		24.12
Supply and install main equipotential earth bonding between incoming mains and other incoming services including earth bonding clamps				
10mm2		Item		81.60
		Carried forward		415.47

	Quantity	Unit	Rate	Total £
		Brought forward		415.47
Supply and install cross-bonding to water services (sinks, boilers, etc.) including earth bonding clamps				
6mm2		Item		41.46
LV SWITCH GEAR AND DISTRIBUTION				
Remove existing consumer unit including the disconnection of the meter and sub-circuits and supply and install new split load consumer unit complete with integral isolator, RCCB, MCBs; all suitable rated				
SP and N; 12 module split-load insulated consumer unit fitted with various MCBs, the remainder fitted with blanks; to backgrounds requiring fixings				
consumer unit	1	each	175.02	175.02
extra for				
32 amp, SP MCB	1	each	10.11	10.11
45 amp, SP MCB	1	each	10.55	10.55
LUMINAIRES AND LAMPS				
Replace and renew existing luminaire including the disconnection of the existing cables and re-connection of the new luminaire				
		Carried forward		652.61

	Quantity	Unit	Rate	Total £
		Brought forward		652.61
White plastic ceiling rose, lamp holder complete with white flexible PVC insulated cable not exceeding 225mm in length; to backgrounds requiring fixings				
pendant	4	each	7.70	30.80
White plastic batten lamp holder complete with skirt; to backgrounds requiring fixings				
ceiling lamp holder	1	each	7.81	7.81
Tungsten luminaire IP44 rated complete with GLS lamp and perspex diffuser; to backgrounds requiring fixings				
60 watt chrome dome, zones 1-3	1	each	25.44	25.44
Fluorescent luminaire complete with lamp and diffuser; to backgrounds requiring fixings				
1500mm long; single tube	1	each	32.78	32.78
ACCESSORIES FOR ELECTRICAL SERVICES				
Replace and renew existing light switch plate including the disconnection of the existing cables and re-connection of the new switch plate; to existing back boxes				
White plastic switch plates				
1 gang; 1 way; single pole	3	each	5.02	15.06
2 gang; 2 way; single pole	1	each	8.47	8.47
		Carried forward		772.97

	Quantity	Unit	Rate	Total £
		Brought forward		772.97
White plastic pull switches including base plate				
1 way; single pole	1	each	10.17	10.17
Replace existing general power accessories including the disconnection of the existing cables and re-connection of the new outlet plate and the installation of flexible PVC insulated earth continuity conductor; to existing back boxes				
White plastic outlet plates				
13 amp; single switched socket outlet	2	each	7.26	14.52
13 amp; twin switched socket outlet	5	each	9.22	46.10
20 amp; double pole indicated switch engraved 'Immersion heater'	1	each	16.37	16.37
13 amp; switch fuse connection unit including outgoing final connections	1	each	12.45	12.45
45 amp; cooker control unit	1	each	19.00	19.00
250 volt grade flush-mounted accessories including back boxes, flexible PVC insulated earth continuity conductor between box and face plate; to backgrounds requiring fixings				
Switch socket outlets				
13 amp; 1 gang	3	each	9.49	28.47
13 amp; 2 gang	4	each	13.26	53.04
		Carried forward		973.09

	Quantity	Unit	Rate	Total £
		Brought forward		973.09
Switch fuse connection unit with indicator including outgoing final connections				
13amp, double pole	2	each	14.93	29.86
Low-level cooker connection unit including final connections				
connection unit	1	each	13.30	13.30
Double pole indicated switch engraved 'Immersion heater'				
20 amp; double pole	1	each	22.86	22.86
Unswitched fuse connection unit including outgoing final connections				
13 amp	1	each	11.08	11.08
250 volt grade surface-mounted accessories including back boxes, to backgrounds requiring fixings				
Indicated pull switch				
45 amp; double pole	1	each	23.23	23.23
240/9 volt grade surface-mounted smoke detector and fire alarm unit complete with rechargeable battery for back-up purposes; hard wired; to backgrounds requiring fixings				
		Carried forward		1,073.42

	Quantity	Unit	Rate	Total £
		Brought forward		1,073.42
Smoke detectors/fire alarm unit complete with associated PVC insulated PVC sheathed cable taken from local lighting circuit				
smoke detector/fire alarm unit	1	each	33.59	33.59
240/9 volt grade surface-mounted carbon monoxide detector complete with pattress; hard wired; to backgrounds requiring fixings				
Carbon monoxide detector unit complete with associated PVC insulated PVC sheathed cable taken from local power circuit, via fuse connection unit				
carbon monoxide detector unit	1	each	54.17	54.17
EXTERNAL LIGHTING				
FITTINGS AND ACCESSORIES				
250 volt grade tungsten luminaire complete with lamp; as manufactured by 'Coughtrie'				
Surface-mounted bulkhead luminaire including conduit sleeve through wall, bushing, final connections and the like; to backgrounds requiring fixings				
'Coughtrie' bulkhead luminaire	1	each	43.80	43.80
		Carried forward		1,204.98

	Quantity	Unit	Rate	Total £
			Brought forward £	1,204.98
Final testing				
Final testing as per IEE Wiring Regulations and production of standard NICEIC documentation per dwelling				
One bedroom, one person flat	1	each	42.00	42.00
Operating and training				
Demonstrate to each tenant the RCCB and MCBs in operation and hand to each tenant a prepared leaflet covering this operation	1	each	6.00	6.00
BUILDER'S WORK IN CONNECTION WITH ELECTRICAL SERVICES				
Cut out in brick or block walls, and make good plaster for				
single socket outlet and/or fuse connection unit	6	each	5.35	32.10
twin socket outlet	4	each	6.50	26.00
low-level cooker connection unit and/or deep pattern socket box	1	each	5.35	5.35
Cut chase in brick or block wall, make good plaster 50mm wide				
cable chase	2	m	1.78	3.56
TOTAL OF UPGRADING OF ELECTRICAL SERVICES TO ONE BEDROOM, ONE PERSON FLAT				1,319.99

	Quantity	Unit	Rate	Total £
ONE BEDROOM, TWO PERSON FLAT				
TESTING				
Electric tests				
100% of circuits		Item		18.00
extra for visual inspection		Item		12.00
extra for complete schedules and report		Item		12.00
REMEDIAL WORK				
LV/HV CABLES AND WIRING				
600/1000 volt grade PVC insulated, PVC sheathed cable; single core				
Remove existing meter tails and renew including liaison with local electricity authority/shipper for accessing meter and re-sealing on completion				
25mm2		Item		27.00
600/1000 volt grade PVC insulated, PVC sheathed; including PVC capping where necessary				
Drawn into voids or chases or clipped to backgrounds; twin and earth cabling				
1.5mm2	15	m	2.28	34.20
2.5mm2	15	m	2.74	41.10
6.0mm2	3	m	7.43	22.29
		Carried forward		166.59

	Quantity	Unit	Rate	Total £
		Brought forward		166.59
Drawn into voids or chases or clipped to backgrounds; three core and earth cabling				
1.5mm2	5	m	4.42	22.10
300/500 volt grade, PVC insulated PVC sheathed heat resistant circular cables; copper stranded conductors; BS 6141; in tails (not exceeding 1 metre long) including termination at both ends				
Remove existing immersion heater cable (in tails) and renew				
3 core; 2.5mm2		Item		29.60
EARTHING AND BONDING COMPONENTS				
Copper earth connection including copper insulated cables and final connection and PVC capping where necessary				
Remove main earth between incoming main and consumer unit and renew				
16mm2		Item		24.12
Supply and install main equipotential earth bonding between incoming mains and other incoming services including earth bonding clamps				
10mm2		Item		81.60
		Carried forward		324.01

	Quantity	Unit	Rate	Total £
		Brought forward		324.01
Supply and install cross-bonding to water services (sinks, boilers, etc.) including earth bonding clamps				
6mm2		Item		41.06
LV SWITCH GEAR AND DISTRIBUTION				
Remove existing consumer unit including the disconnection of the meter and sub-circuits and supply and install new split-load consumer unit complete with integral isolator, RCCB, MCBs; all suitable rated				
SP and N; 12 module split-load insulated consumer unit fitted with various MCBs, the remainder fitted with blanks; to backgrounds requiring fixings				
consumer unit	1	each	175.02	175.02
extra for				
6 amp, SP MCB	1	each	10.55	10.55
32 amp, SP MCB	1	each	10.11	10.11
45 amp, SP MCB	1	each	10.55	10.55
bell transformer	1	each	23.45	23.45
LUMINAIRES AND LAMPS				
Replace and renew existing luminaire including the disconnection of the existing cables and re-connection of the new luminaire				
		Carried forward		594.75

	Quantity	Unit	Rate	Total £
		Brought forward		594.75
White plastic ceiling rose, lamp holder complete with white flexible PVC insulated cable not exceeding 225mm in length; to backgrounds requiring fixings				
pendant	3	each	7.7	23.10
Recessed mains voltage luminaires IP44 rated complete with opal lens and lamp; fitted into recess formed by others				
60 watt tungsten downlighter, zones 1-3	3	each	18.56	55.68
Fluorescent luminaire complete with lamp and perspex diffuser; to backgrounds requiring fixings				
1500mm long; single tube	1	each	32.78	32.78
ACCESSORIES FOR ELECTRICAL SERVICES				
Replace and renew existing light switch plate including the disconnection of the existing cables and re-connection of the new switch plate; to existing back boxes				
White plastic switch plates				
1 gang; 1 way; single pole	3	each	5.02	15.06
2 gang; 2 way; single pole	3	each	8.47	25.41
		Carried forward		746.78

	Quantity	Unit	Rate	Total £
		Brought forward		746.78
White plastic pull switches including base plate				
1 way; single pole	1	each	10.17	10.17
Replace existing general power accessories including the disconnection of the existing cables and re-connection of the new outlet plate and the installation of flexible PVC insulated earth continuity conductor; to existing back boxes				
White plastic outlet plates				
13 amp; single switched socket outlet	2	each	7.26	14.52
13 amp; twin switched socket outlet	7	each	9.22	64.54
20 amp; double pole indicated switch engraved 'Immersion heater'	1	each	16.37	16.37
13 amp; switch fuse connection unit including outgoing final connections	1	each	12.45	12.45
45 amp; cooker control unit	1	each	19.00	19.00
250 volt grade flush-mounted accessories including back boxes, flexible PVC insulated earth continuity conductor between box and face plate; to backgrounds requiring fixings				
Switch socket outlets				
13 amp; 1 gang	4	each	9.49	37.96
13 amp; 2 gang	3	each	13.26	39.78
		Carried forward		961.57

	Quantity	Unit	Rate	Total £
		Brought forward		961.57
Switch fuse indicated unit including outgoing final connections				
13amp; double pole	3	each	14.93	44.79
Low-level cooker connection unit including final connections				
connection unit	1	each	13.30	13.30
Double pole indicated switch engraved 'Immersion heater'				
20 amp; double pole	1	each	22.86	22.86
Unswitched fuse connection unit including outgoing final connections				
13 amp	1	each	11.08	11.08
Sundry works				
Door chimes and bell push including associated wiring and protection; to backgrounds requiring fixings				
door bell system	1	each	36.98	36.98
250 volt grade surface-mounted accessories including back boxes, to backgrounds requiring fixings				
Indicated pull switch				
45 amp; double pole	1	each	23.23	23.23
		Carried forward		1,113.81

	Quantity	Unit	Rate	Total £
		Brought forward		1,113.81
240/9 volt grade surface-mounted smoke detector and fire alarm unit complete with rechargeable battery for back-up purposes; hard wired; to backgrounds requiring fixings				
Smoke detectors/fire alarm unit complete with associated PVC insulated PVC sheathed cable taken from local lighting circuit				
smoke detector/fire alarm unit	1	each	33.59	33.59
240/9 volt grade surface-mounted carbon monoxide detector and fire alarm unit complete with re-chargeable battery for back-up purposes; hard wired; to backgrounds requiring fixings				
Carbon monoxide detector unit complete with associated PVC insulated PVC sheathed cable taken from local power cirvuit via fuse connection unit				
carbon monoxide detector unit	1	each	54.17	54.17
EXTERNAL LIGHTING				
FITTINGS AND ACCESSORIES				
250 volt grade tungsten luminaire complete with lamp; as manufactured by 'Coughtrie'				
Surface mounted bulkhead luminaire including conduit sleeve through wall, bushing, final connections and the like; to backgrounds requiring fixings				
'Coughtrie' bulkhead luminaire	1	each	43.80	43.80
		Carried forward		1,245.37

	Quantity	Unit	Rate	Total £
		Brought forward		1,245.37
Final testing				
Final testing as per IEE Wiring Regulations and production of standard NICEIC documentation per dwelling				
One bedroom, two person flat	1	each	48.00	48.00
Operating and training				
Demonstrate to each tenant the RCCB and MCBs in operation and hand to each tenant a prepared leaflet covering this operation	1	each	6.00	6.00
BUILDER'S WORK IN CONNECTION WITH ELECTRICAL SERVICES				
Cut out in brick or block walls, and make good plaster for				
single socket outlet and/or fuse connection unit	8	each	5.35	42.80
twin socket outlet	3	each	6.50	19.50
low-level cooker connection unit and/or deep pattern socket box	1	each	5.35	5.35
Cut chase in brick or block wall, make good plaster 50mm wide				
cable chase	4	m	1.78	7.12
TOTAL OF UPGRADING OF ELECTRICAL SERVICES TO ONE BEDROOM, TWO PERSON FLAT				1,374.14

	Quantity	Unit	Rate	Total £
TWO BEDROOM, THREE PERSON HOUSE				
TESTING				
Electric tests				
100% of circuits		Item		24.00
extra for visual inspection		Item		15.00
extra for complete schedules and report		Item		15.00
REMEDIAL WORK				
LV/HV CABLES AND WIRING				
600/1000 volt grade PVC insulated, PVC sheathed cable, single core				
Remove existing meter tails and renew including liaison with local electricity authority/shipper for accessing meter and re-sealing on completion				
25mm2		Item		27.00
600/1000 volt grade PVC insulated, PVC sheathed; including PVC capping where necessary				
Drawn into voids or chases or clipped to backgrounds; twin and earth cabling				
1.5mm2	10	m	2.28	22.80
2.5mm2	25	m	2.74	68.50
6.0mm2	2	m	7.43	14.86
		Carried forward		187.16

	Quantity	Unit	Rate	Total £
		Brought forward		187.16
Drawn into voids or chases or clipped to backgrounds; three core and earth cabling				
1.5mm2	15	m	4.42	66.30
300/500 volt grade, PVC insulated PVC sheathed heat resistant circular cables; copper stranded conductors; BS 6141; in tails (not exceeding 1 metre long) including termination at both ends				
Remove existing immersion heater cable (in tails) and renew				
3 core; 2.5mm2		Item		29.60
EARTHING AND BONDING COMPONENTS				
Copper earth connection including copper insulated cables and final connection and PVC capping where necessary				
Remove main earth between incoming main and consumer unit and renew				
16mm2		Item		24.12
Supply and install main equipotential earth bonding between incoming mains and other incoming services including earth bonding clamps				
10mm2		Item		81.60
		Carried forward		388.78

	Quantity	Unit	Rate	Total £
		Brought forward		388.78
Supply and install cross-bonding to water services (sinks, boilers, etc.) including earth bonding clamps 6mm2		Item		41.46
LV SWITCH GEAR AND DISTRIBUTION				
Remove existing consumer unit including the disconnection of the meter and sub-circuits and supply and install new split load consumer unit complete with integral isolator, RCCB, MCBs; all suitable rated				
SP and N; 12 module split-load insulated consumer unit fitted with various MCBs, the remainder fitted with blanks; to backgrounds requiring fixings				
consumer unit	1	each	175.02	175.02
extra for				
6 amp, SP MCB	1	each	10.55	10.55
16 amp, SP MCB	1	each	10.11	10.11
32 amp, SP MCB	1	each	10.11	10.11
45 amp, SP MCB	1	each	10.55	10.55
LUMINAIRES AND LAMPS				
Replace and renew existing luminaire including the disconnection of the existing cables and re-connection of the new luminaire				
		Carried forward		646.58

	Quantity	Unit	Rate	Total £
		Brought forward		646.58
White plastic ceiling rose, lamp holder complete with white flexible PVC insulated cable not exceeding 225mm in length; to backgrounds requiring fixings				
pendant	6	each	7.70	46.20
Tungsten luminaire IP44 rated complete with GLS lamp and perspex diffuser; to backgrounds requiring fixings				
60 watt chrome dome, zones 1-3	1	each	25.44	25.44
Fluorescent luminaire complete with lamp and perspex diffuser; to backgrounds requiring fixings				
1500mm long; single tube	1	each	32.78	32.78
ACCESSORIES FOR ELECTRICAL SERVICES				
Replace and renew existing light switch plate including the disconnection of the existing cables and re-connection of the new switch plate; to existing back boxes				
White plastic switch plates				
1 gang; 1 way; single pole	3	each	5.02	15.06
2 gang; 2 way; single pole	1	each	8.47	8.47
3 gang; 2 way; single pole	1	each	14.80	14.80
1 gang; 2 way; single pole	3	each	5.80	17.40
		Carried forward		806.73

	Quantity	Unit	Rate	Total £
		Brought forward		806.73
White plastic pull switches including base plate				
1 way; single pole	1	each	10.17	10.17
Replace existing general power accessories including the disconnection of the existing cables and re-connection of the new outlet plate and the installation of flexible PVC insulated earth continuity conductor; to existing back boxes				
White plastic outlet plates				
13 amp; single switched socket outlet	3	each	7.26	21.78
13 amp; twin switched socket outlet	6	each	9.22	55.32
20 amp; double pole indicated switch engraved 'Immersion heater'	2	each	16.37	32.74
13 amp; switch fuse connection unit including outgoing final connections	1	each	12.45	12.45
45 amp; cooker control unit	1	each	19.00	19.00
250 volt grade flush-mounted accessories including back boxes, flexible PVC insulated earth continuity conductor between box and face plate; to backgrounds requiring fixings				
Switch socket outlets				
13 amp; 1 gang	4	each	9.49	37.96
13 amp; 2 gang	5	each	13.26	66.30
		Carried forward		1,062.45

	Quantity	Unit	Rate	Total £
		Brought forward		1,062.45
Switch fuse indicated unit including outgoing final connections				
13amp; double pole	3	each	14.93	44.79
Low-level cooker connection unit including final connections				
connection unit	1	each	13.30	13.30
Double pole indicated switch engraved 'Immersion heater'				
20 amp; double pole	1	each	22.86	22.86
Unswitched fuse connection unit including outgoing final connections				
13 amp	1	each	11.08	11.08
250 volt grade surface-mounted accessories including back boxes, to backgrounds requiring fixings				
Indicated pull switch with indicator				
45 amp; double pole	1	each	23.23	23.23
240/9 volt grade surface-mounted smoke detector and fire alarm unit complete with rechargeable battery for back-up purposes; hard wired; to backgrounds requiring fixings				
		Carried forward		1,177.71

	Quantity	Unit	Rate	Total £
		Brought forward		1,177.71
Smoke detectors/fire alarm unit complete with associated PVC insulated PVC sheathed cable taken from local lighting circuit				
smoke detector/fire alarm unit	2	each	33.59	67.18
240/9 volt grade surface-mounted carbon monoxide detector and fire alarm unit complete with re-chargeable battery for back-up purposes; hard wired; to backgrounds requiring fixings				
Carbon monoxide detector unit complete with associated PVC insulated PVC sheathed cable taken from local power circuit via fuse connection unit				
carbon monoxide detector unit	1	each	54.17	54.17
EXTERNAL LIGHTING				
FITTINGS AND ACCESSORIES				
250 volt grade tungsten luminaire complete with lamp; as manufactured by 'Coughtrie'				
Surface-mounted bulkhead luminaire including conduit sleeve through wall, bushing, final connections and the like; to backgrounds requiring fixings				
'Coughtrie' bulkhead luminaire	1	each	43.80	43.80
		Carried forward		1,342.86

	Quantity	Unit	Rate	Total £
		Brought forward		1,342.86
Final testing				
Final testing as per IEE Wiring Regulations and production of standard NICEIC documentation per dwelling				
Two bedroom, three person house	1	each	54.00	54.00
Operating and training				
Demonstrate to each tenant the RCCB and MCBs in operation and hand to each tenant a prepared leaflet covering this operation	1	each	6.00	6.00
BUILDER'S WORK IN CONNECTION WITH ELECTRICAL SERVICES				
Cut out in brick or block walls, and make good plaster for				
single socket outlet and/or fuse connection unit	9	each	5.35	48.15
twin socket outlet	5	each	6.50	32.50
low-level cooker connection unit and/or deep pattern socket box	1	each	5.35	5.35
Cut chase in brick or block wall, make good plaster 50mm wide				
cable chase	10	m	1.78	17.80
TOTAL OF UPGRADING OF ELECTRICAL SERVICES TO TWO BEDROOM, THREE PERSON HOUSE				1,506.66

	Quantity	Unit	Rate	Total £
TWO BEDROOM, FOUR PERSON HOUSE				
TESTING				
Electric tests				
100% of circuits		Item		24.00
extra for visual inspection		Item		15.00
extra for complete schedules and report		Item		15.00
REMEDIAL WORK				
LV/HV CABLES AND WIRING				
600/1000 volt grade PVC insulated, PVC sheathed cable, single core				
Remove existing meter tails and renew including liaison with local electricity authority/shipper for accessing meter and re-sealing on completion				
25mm2		Item		27.00
600/1000 volt grade PVC insulated, PVC sheathed; including PVC capping where necessary				
Drawn into voids or chases or clipped to backgrounds; twin and earth cabling				
1.5mm2	15	m	2.28	34.20
2.5mm2	35	m	2.74	95.90
6.0mm2	20	m	7.43	148.60
		Carried forward		359.70

	Quantity	Unit	Rate	Total £
		Brought forward		359.70
Drawn into voids or chases or clipped to backgrounds; three core and earth cabling				
1.5mm2	15	m	4.42	66.30
300/500 volt grade, PVC insulated PVC sheathed heat resistant circular cables; copper stranded conductors; BS 6141; in tails (not exceeding 1 metre long) including termination at both ends				
Remove existing immersion heater cable (in tails) and renew				
3 core; 2.5mm2		Item		29.60
EARTHING AND BONDING COMPONENTS				
Copper earth connection including copper insulated cables and final connection and PVC capping where necessary				
Remove main earth between incoming main and consumer unit and renew				
16mm2		Item		24.12
Supply and install main equipotential earth bonding between incoming mains and other incoming services including earth bonding clamps				
10mm2		Item		81.60
		Carried forward		561.32

	Quantity	Unit	Rate	Total £
		Brought forward		561.32
Supply and install cross-bonding to water services (sinks, boilers, etc.) including earth bonding clamps				
6mm2		Item		41.46
LV SWITCH GEAR AND DISTRIBUTION				
Remove existing consumer unit including the disconnection of the meter and sub-circuits and supply and install new split-load consumer unit complete with integral isolator, RCCB, MCBs; all suitable rated				
SP and N; 15 module split-load insulated consumer unit fitted with various MCBs, the remainder fitted with blanks; to backgrounds requiring fixings				
consumer unit	1	each	183.31	183.31
extra for				
6 amp, SP MCB	1	each	10.55	10.55
16 amp, SP MCB	1	each	10.11	10.11
32 amp, SP MCB	2	each	10.11	20.22
45 amp, SP MCB	1	each	10.55	10.55
80 amp; 100mAmp RCCB	1	each	72.89	72.89
LUMINAIRES AND LAMPS				
Replace and renew existing luminaire including the disconnection of the existing cables and re-connection of the new luminaire				
		Carried forward		910.41

	Quantity	Unit	Rate	Total £
		Brought forward		910.41
White plastic ceiling rose, lamp holder complete with white flexible PVC insulated cable not exceeding 225mm in length; to backgrounds requiring fixings				
pendant	4	each	7.70	30.80
White plastic batten lamp holder complete with skirt; to backgrounds requiring fixings				
ceiling lamp holder	1	each	7.81	7.81
Recessed mains voltage luminaires IP44 rated complete with opal lens and lamp; fitted into recess formed by others				
60 watt tungsten downlighter	4	each	18.56	74.24
Disconnect, remove, test and re-install tenant's own decorative metal luminaire				
tenant's luminaire (wall)	2	each	17.57	35.14
ACCESSORIES FOR ELECTRICAL SERVICES				
Replace and renew existing light switch plate including the disconnection of the existing cables and re-connection of the new switch plate; to existing back boxes				
White plastic switch plates				
1 gang; 1 way; single pole	4	each	5.02	20.08
2 gang; 2 way; single pole	1	each	8.47	8.47
3 gang; 2 way; single pole	1	each	14.80	14.80
1 gang; 2 way; single pole	3	each	5.80	17.40
		Carried forward		1,119.15

	Quantity	Unit	Rate	Total £
		Brought forward		1,119.15
White plastic pull switches including base plate				
1 way; single pole	1	each	10.17	10.17
Replace existing general power accessories including the disconnection of the existing cables and re-connection of the new outlet plate and the installation of flexible PVC insulated earth continuity conductor; to existing back boxes				
White plastic outlet plates				
13 amp; single switched socket outlet	4	each	7.26	29.04
13 amp; twin switched socket outlet	2	each	9.22	18.44
20 amp; double pole indicated switch engraved 'Immersion heater'	1	each	16.37	16.37
13 amp; switch fuse connection unit including outgoing final connections	1	each	12.45	12.45
45 amp; cooker control unit	1	each	19.00	19.00
low-level cooker connection unit	1	each	12.05	12.05
250 volt grade flush-mounted accessories including back boxes, flexible PVC insulated earth continuity conductor between box and face plate; to backgrounds requiring fixings				
Switch socket outlets				
13 amp; 1 gang	2	each	9.49	18.98
13 amp; 2 gang	10	each	13.26	132.60
		Carried forward		1,388.25

	Quantity	Unit	Rate	Total £
		Brought forward		1,388.25
Switch fuse indicated unit including outgoing final connections				
13 amp; double pole	5	each	14.93	74.65
Double pole indicated switch engraved 'Immersion heater'				
20 amp; double pole	1	each	22.86	22.86
Unswitched fuse connection unit including outgoing final connections				
13 amp	1	each	11.08	11.08
250 volt grade surface mounted accessories including back boxes; to backgrounds requiring fixings				
Indicated pull switch				
45 amp, double pole	1	each	23.23	23.23
Install tenant's own electric shower				
Connect electric shower to existing cold water supply adjacent unit; to backgrounds requiring fixings				
electric shower	1	each	38.69	38.69
240/9 volt grade surface-mounted smoke detector and fire alarm unit complete with rechargeable battery for back-up purposes; hard wired; to backgrounds requiring fixings				
		Carried forward		1,558.76

	Quantity	Unit	Rate	Total £
		Brought forward		1,558.76
Smoke detectors/fire alarm unit complete with associated PVC insulated PVC sheathed cable taken from local lighting circuit				
smoke detector/fire alarm unit	2	each	33.59	67.18
EXTERNAL LIGHTING				
FITTINGS AND ACCESSORIES				
250 volt grade tungsten luminaire complete with lamp; as manufactured by 'Coughtrie'				
Surface-mounted bulkhead luminaire including conduit sleeve through wall, bushing, final connections and the like; to backgrounds requiring fixings				
'Coughtrie' bulkhead luminaire	2	each	43.80	87.60
Final testing				
Final testing as per IEE Wiring Regulations and production of standard NICEIC documentation per dwelling				
Two bedroom, four person house	1	each	54.00	54.00
Operating and training				
Demonstrate to each tenant the RCCB and MCBs in operation and hand to each tenant a prepared leaflet covering this operation	1	each	6.00	6.00
		Carried forward		1,773.54

	Quantity	Unit	Rate	Total £
		Brought forward		1,773.54
BUILDER'S WORK IN CONNECTION WITH ELECTRICAL SERVICES				
Cut out in brick or block walls, and make good plaster for				
single socket outlet and/or fuse connection unit	9	each	5.35	48.15
twin socket outlet	10	each	6.50	65.00
Cut chase in brick or block wall, make good plaster 50mm wide				
cable chase	30	m	1.78	53.40
Floor boards, tongued and grooved				
lift softwood flooring; using hand tools	15	m	6.25	93.75
relay softwood flooring; fixing with screws	15	m	2.07	31.05
TOTAL OF UPGRADING OF ELECTRICAL SERVICES TO TWO BEDROOM, FOUR PERSON HOUSE				2,064.89

	Quantity	Unit	Rate	Total £
THREE BEDROOM, FOUR PERSON HOUSE				
TESTING				
Electric tests				
100% of circuits		Item		24.00
extra for visual inspection		Item		15.00
extra for complete schedules and report		Item		15.00
REMEDIAL WORK				
LV/HV CABLES AND WIRING				
600/1000 volt grade PVC insulated, PVC sheathed cable, single core				
Remove existing meter tails and renew including liaison with local electricity authority/shipper for accessing meter and re-sealing on completion				
25mm2		Item		27.00
600/1000 volt grade PVC insulated, PVC sheathed; including PVC capping where necessary				
Drawn into voids or chases or clipped to backgrounds; twin and earth cabling				
1.5mm2	20	m	2.28	45.60
2.5mm2	48	m	2.74	131.52
6.0mm2	5	m	7.43	37.15
			Carried forward	295.27

	Quantity	Unit	Rate	Total £
		Brought forward		295.27
Drawn into voids or chases or clipped to backgrounds; three core and earth cabling				
1.5mm2	15	m	4.42	66.30
300/500 volt grade, PVC insulated PVC sheathed heat resistant circular cables; copper stranded conductors; BS 6141; in tails (not exceeding 1 metre long) including termination at both ends				
Remove existing immersion heater and boiler cable (in tails) and renew				
3 core; 2.5mm2		Item		29.60
EARTHING AND BONDING COMPONENTS				
Copper earth connection including copper insulated cables and final connection and PVC capping where necessary				
Remove main earth between incoming main and consumer unit and renew				
16mm2		Item		24.12
Supply and install main equipotential earth bonding between incoming mains and other incoming services including earth bonding clamps				
10mm2		Item		81.60
		Carried forward		496.89

	Quantity	Unit	Rate	Total £
		Brought forward		496.89
Supply and install cross-bonding to water services (sinks, boilers, etc.) including earth bonding clamps				
6mm2		Item		41.46
LV SWITCH GEAR AND DISTRIBUTION				
Remove existing consumer unit including the disconnection of the meter and sub-circuits and supply and install new split-load consumer unit complete with integral isolator, RCCB, MCBs; all suitable rated				
SP and N; 15 module split-load insulated consumer unit fitted with various MCBs, the remainder fitted with blanks; to backgrounds requiring fixings				
consumer unit	1	each	183.31	183.31
extra for				
6 amp, SP MCB	1	each	10.55	10.55
32 amp, SP MCB	1	each	10.11	10.11
45 amp, SP MCB	1	each	10.55	10.55
80 amp; 100mA, RCCB	1	each	72.89	72.89
bell transformer	1	each	23.45	23.45
LUMINAIRES AND LAMPS				
Replace and renew existing luminaire including the disconnection of the existing cables and re-connection of the new luminaire				
		Carried forward		849.21

	Quantity	Unit	Rate	Total £
		Brought forward		849.21
White plastic ceiling rose, lamp holder complete with white flexible PVC insulated cable not exceeding 225mm in length; to backgrounds requiring fixings				
pendant	5	each	7.70	38.50
Tungsten luminaire IP44 rated complete with GLS lamp[and diffuser; to backgrounds requiring requiring fixings				
60 watt chrome dome, zones 1-3	1	each	25.44	25.44
Fluorescent luminaire complete with lamp and perspex diffuser; to backgrounds fixings				
1500mm long; single tube	1	each	32.78	32.78
Install tenant's own wall-mounted luminaire; to backgrounds requiring fixings				
tungsten wall lights	2	each	11.44	22.88
ACCESSORIES FOR ELECTRICAL SERVICES				
Replace and renew existing light switch plate including the disconnection of the existing cables and re-connection of the new switch plate; to existing back boxes				
		Carried forward		968.81

	Quantity	Unit	Rate	Total £
		Brought forward		968.81
White plastic switch plates				
1 gang; 1 way; single pole	5	each	5.02	25.10
2 gang; 2 way; single pole	1	each	8.47	8.47
3 gang; 2 way; single pole	1	each	14.80	14.80
1 gang; 2 way; single pole	1	each	5.80	5.80
White plastic pull switches including base plate				
1 way; single pole	1	each	10.17	10.17
Replace existing general power accessories including the disconnection of the existing cables and re-connection the new outlet plate and the installation of flexible PVC insulated earth continuity conductor; to excisting back boxes				
White plastic outlet plates				
13 amp; single switched socket outlet	4	each	7.26	29.04
13 amp; twin switched socket outlet	9	each	9.22	82.98
20 amp; double pole indicated switch engraved 'Immersion heater'	2	each	16.37	32.74
13 amp; switch fuse connection unit including outgoing final connections	1	each	12.45	12.45
45 amp; cooker control unit	1	each	19.00	19.00
low-level cooker connection unit	1	each	12.05	12.05
250 volt grade flush-mounted accessories including back boxes, flexible PVC insulated earth continuity conductor between box and face plate; to backgrounds requiring fixings				
		Carried forward		1,221.41

	Quantity	Unit	Rate	Total £
		Brought forward		1,221.41
Switch socket outlets				
13 amp; 1 gang	2	each	9.49	18.98
13 amp; 2 gang	5	each	13.26	66.30
Switch fuse indicated unit including outgoing final connections				
13 amp; double pole	4	each	14.93	59.72
Double pole indicated switch engraved 'Immersion heater'				
20 amp; double pole	1	each	22.86	22.86
Unswitched fuse connection unit including outgoing final connections				
13 amp	1	each	11.08	11.08
Sundries				
Door chimes and bell push including associated wiring and protection; to backgrounds requiring fixings				
door bell system	1	each	36.98	36.98
240/9 volt grade surface-mounted smoke detector and fire alarm unit complete with rechargeable battery for back-up purposes; hard wired; to backgrounds requiring fixings				
		Carried forward		1,437.33

	Quantity	Unit	Rate	Total £
		Brought forward		1,437.33
Smoke detectors/fire alarm unit complete with associated PVC insulated PVC sheathed cable taken from local lighting circuit				
smoke detector/fire alarm unit	2	each	33.59	67.18
240/9 volt grade surface-mounted carbon monoxide detector and fire alarm unit complete with re-chargeable battery for back-up purposes; hard wired; to backgrounds requiring fixings				
Carbon monoxide detector unit complete with associated PVC insulated PVC sheathed cable taken from local power circuit via fuse connection unit				
carbon monoxide detector unit	1	each	54.17	54.17
EXTERNAL LIGHTING				
FITTINGS AND ACCESSORIES				
250 volt grade tungsten luminaire complete with lamp; as manufactured by 'Coughtrie'				
Surface-mounted bulkhead luminaire including conduit sleeve through wall, bushing, final connections and the like; to backgrounds requiring fixings				
'Coughtrie' bulkhead luminaire	2	each	43.80	87.60
		Carried forward		1,646.28

	Quantity	Unit	Rate	Total £
		Brought forward		1,646.28
Final testing				
Final testing as per IEE Wiring Regulations and production of standard NICEIC documentation per dwelling				
Three bedroom, four person house	1	each	54.00	54.00
Operating and training				
Demonstrate to each tenant the RCCB and MCBs in operation and hand to each tenant a prepared leaflet covering this operation	1	each	6.00	6.00
BUILDER'S WORK IN CONNECTION WITH ELECTRICAL SERVICES				
Cut out in brick or block walls, and make good plaster for				
single socket outlet and/or fuse connection unit	7	each	5.35	37.45
twin socket outlet	8	each	6.50	52.00
low-level cooker connection unit and/or deep pattern socket box	1	each	5.35	5.35
Cut chase in brick or block wall, make good plaster 50mm wide				
cable chase	35	m	1.78	62.30
		Carried forward		1,863.38

	Quantity	Unit	Rate	Total £
		Brought forward		1,863.38
Floor boards, tongued and grooved				
lift softwood flooring; using hand tools	15	m	6.25	93.75
relay softwood flooring; fixing with screws	15	m	2.07	31.05
TOTAL OF UPGRADING OF ELECTRICAL SERVICES TO THREE BEDROOM, FOUR PERSON HOUSE				1,988.18

	Quantity	Unit	Rate	Total £
THREE BEDROOM FIVE PERSON HOUSE				
TESTING				
Electric tests				
100% of circuits		Item		30.00
extra for visual inspection		Item		18.00
extra for complete schedules and report		Item		18.00
REMEDIAL WORK				
LV/HV CABLES AND WIRING				
600/1000 volt grade PVC insulated, PVC sheathed cable, single core				
Remove existing meter tails and renew including liaison with local electricity authority/shipper for accessing meter and re-sealing on completion				
25mm2		Item		27.00
600/1000 volt grade PVC insulated, PVC sheathed; including PVC capping where necessary				
Drawn into voids or chases or clipped to backgrounds; twin and earth cabling				
1.5mm2	18	m	2.28	41.04
2.5mm2	30	m	2.74	82.20
6.0mm2	20	m	7.43	148.60
		Carried forward		364.84

	Quantity	Unit	Rate	Total £
		Brought forward		364.84
Drawn into voids or chases or clipped to backgrounds; three core and earth cabling				
1.5mm2	15	m	4.42	66.30
300/500 volt grade, PVC insulated PVC sheathed heat resistant circular cables; copper stranded conductors; BS 6141; in tails (not exceeding 1 metre long) including termination at both ends				
Remove existing immersion heater and boiler cables (in tails) and renew				
3 core; 2.5mm2		Item		29.60
EARTHING AND BONDING COMPONENTS				
Copper earth connection including copper insulated cables and final connection and PVC capping where necessary				
Remove main earth between incoming main and consumer unit and renew				
16mm2		Item		24.12
Supply and install main equipotential earth bonding between incoming mains and other incoming services including earth bonding clamps				
10mm2		Item		81.60
		Carried forward		566.46

	Quantity	Unit	Rate	Total £
		Brought forward		566.46
Supply and install cross-bonding to water services (sinks, boilers, etc.) including earth bonding clamps				
6mm2		Item		41.46
LV SWITCH GEAR AND DISTRIBUTION				
Remove existing consumer unit including the disconnection of the meter and sub-circuits and supply and install new split-load consumer unit complete with integral isolator, RCCB, MCBs; all suitable rated				
SP and N; 17 module split-load insulated consumer unit fitted with various MCBs, the remainder fitted with blanks; to backgrounds requiring fixings				
consumer unit	1	each	195.51	195.51
extra for				
6 amp, SP MCB	2	each	10.55	21.10
16 amp, SP MCB	1	each	10.11	10.11
32 amp, SP MCB	2	each	10.11	20.22
45 amp, SP MCB	1	each	10.55	10.55
bell transformer	1	each	23.45	23.45
LUMINAIRES AND LAMPS				
Replace and renew existing luminaire including the disconnection of the existing cables and re-connection of the new luminaire				
		Carried forward		888.86

	Quantity	Unit	Rate	Total £
		Brought forward		888.86
White plastic ceiling rose, lamp holder complete with white flexible PVC insulated cable not exceeding 225mm in length; to backgrounds requiring fixings				
pendant	4	each	7.70	30.80
White plastic batten lamp holder complete with skirt; to backgrounds requiring fixings				
ceiling lamp holder	1	each	7.81	7.81
Tungsten luminaire IP44 rated complete with GLS lamp and diffuser; to backgrounds requiring requiring fixings				
60 watt chrome dome, zones 1-3	2	each	25.44	50.88
Fluorescent luminaire complete with lamp and perspex diffuser; to backgrounds requiring fixings				
1500mm long; single tube	1	each	32.78	32.78
Disconnect, remove, test and re-install tenant's own decorative metal luminaire				
tenant's luminaire (ceiling)	2	each	17.57	35.14
Install tenant's own wall-mounted luminaire; to backgrounds requiring fixings				
tungsten wall lights	2	each	11.44	22.88
		Carried forward		1,069.15

	Quantity	Unit	Rate	Total £
		Brought forward		1,069.15
ACCESSORIES FOR ELECTRICAL SERVICES				
Replace and renew existing light switch plate including the disconnection of the existing cables and re-connection of the new switch plate; to existing back boxes				
White plastic switch plates				
1 gang; 1 way; single pole	6	each	5.02	30.12
2 gang; 2 way; single pole	1	each	8.47	8.47
1 gang; 2 way; single pole	5	each	5.80	29.00
White plastic pull switches including base plate				
1 way; single pole	1	each	10.17	10.17
Replace existing general power accessories including the disconnection of the existing cables and re-connection of the new outlet plate and the installation of flexible PVC insulated earth continuity conductor; to existing back boxes				
White plastic outlet plates				
13 amp; single switched socket outlet	2	each	7.26	14.52
13 amp; twin switched socket outlet	12	each	9.22	110.64
20 amp; double pole indicated switch engraved 'Immersion heater'	2	each	16.37	32.74
13 amp; switch fuse connection unit including outgoing final connections	1	each	12.45	12.45
45 amp; cooker control unit	1	each	19.00	19.00
		Carried forward		1,336.26

	Quantity	Unit	Rate	Total £
		Brought forward		1,336.26
250 volt grade flush-mounted accessories including back boxes, flexible PVC insulated earth continuity conductor between box and face plate; to backgrounds requiring fixings				
Switch socket outlets				
13 amp; 1 gang	2	each	9.49	18.98
13 amp; 2 gang	3	each	13.26	39.78
Switch fuse indicated unit including outgoing final connections				
13 amp; double pole	4	each	14.93	59.72
Low-level cooker connection unit including final connections				
connection unit	1	each	13.30	13.30
Double pole indicated switch engraved 'Immersion heater'				
20 amp; double pole	1	each	22.86	22.86
Unswitched fuse connection unit including outgoing final connections				
13 amp	1	each	11.08	11.08
		Carried forward		1,501.98

	Quantity	Unit	Rate	Total £
		Brought forward		1,501.98
Sundry works				
Door chimes and bell push including associated wiring and protection; to backgrounds requiring fixings				
door bell system	1	each	36.98	36.98
240/9 volt grade surface-mounted smoke detector and fire alarm unit complete with rechargeable battery for back-up purposes; hard wired; to backgrounds requiring fixings				
Smoke detectors/fire alarm unit complete with associated PVC insulated PVC sheathed cable taken from local lighting circuit				
smoke detector/fire alarm unit	2	each	33.59	67.18
240/9 volt grade surface-mounted carbon monoxide detector and fire alarm unit complete with re-chargeable battery for back-up purposes; hard wired; to backgrounds requiring fixings				
Carbon monoxide detector unit complete with associated PVC insulated PVC sheathed cable taken from local power circuit via fuse connection unit				
carbon monoxide detector unit	1	each	54.17	54.17
		Carried forward		1,660.31

	Quantity	Unit	Rate	Total £
		Brought forward		1,660.31
EXTERNAL LIGHTING				
FITTINGS AND ACCESSORIES				
250 volt grade tungsten luminaire complete with lamp; as manufactured by 'Coughtrie'				
Surface-mounted bulkhead luminaire including conduit sleeve through wall, bushing, final connections and the like; to backgrounds requiring fixings				
'Coughtrie' bulkhead luminaire	2	each	43.80	87.60
Final testing				
Final testing as per IEE Wiring Regulations and production of standard NICEIC documentation per dwelling				
Three bedroom, five person house	1	each	60.00	60.00
Operating and training				
Demonstrate to each tenant the RCCB and MCBs in operation and hand to each tenant a prepared leaflet covering this operation	1	each	6.00	6.00
		Carried forward		1,813.91

	Quantity	Unit	Rate	Total £
		Brought forward		1,813.91
BUILDER'S WORK IN CONNECTION WITH ELECTRICAL SERVICES				
Cut out in brick or block walls, and make good plaster for				
single socket outlet and/or fuse connection unit	6	each	5.35	32.10
twin socket outlet	3	each	6.50	19.50
low-level cooker connection unit and/or deep pattern socket box	2	each	5.35	10.70
Cut chase in brick or block wall, make good plaster 50mm wide				
cable chase	40	m	1.78	71.20
Floor boards, butt jointed				
lift softwood flooring; using hand tools	10	m	3.86	38.60
relay softwood flooring; fixing with screws	10	m	2.07	20.70
TOTAL OF UPGRADING OF ELECTRICAL SERVICES TO THREE BEDROOM, FIVE PERSON HOUSE				2,006.71

	Quantity	Unit	Rate	Total £
THREE BEDROOM, SIX PERSON HOUSE				
TESTING				
Electric tests				
100% of circuits		Item		30.00
extra for visual inspection		Item		18.00
extra for complete schedules and report		Item		18.00
REMEDIAL WORK				
LV/HV CABLES AND WIRING				
600/1000 volt grade PVC insulated, PVC sheathed cable, single core				
Remove existing meter tails and renew including liaison with local electricity authority/shipper for accessing meter and re-sealing on completion				
25mm2		Item		27.00
600/1000 volt grade PVC insulated, PVC sheathed; including PVC capping where necessary				
Drawn into voids or chases or clipped to backgrounds; twin and earth cabling				
1.5mm2	18	m	2.28	41.04
2.5mm2	45	m	2.74	123.30
6.0mm2	20	m	7.43	148.60
		Carried forward		405.94

	Quantity	Unit	Rate	Total £
		Brought forward		405.94
Drawn into voids or chases or clipped to backgrounds; three core and earth cabling				
1.5mm2	18	m	4.42	79.56
300/500 volt grade, PVC insulated PVC sheathed heat resistant circular cables; copper stranded conductors; BS 6141; in tails (not exceeding 1 metre long) including termination at both ends				
Remove existing immersion heater and boiler cables (in tails) and renew				
3 core; 2.5mm2		Item		29.60
EARTHING AND BONDING COMPONENTS				
Copper earth connection including copper insulated cables and final connection and PVC capping where necessary				
Remove main earth between incoming main and consumer unit and renew				
16mm2		Item		24.12
Supply and install main equipotential earth bonding between incoming mains and other incoming services including earth bonding clamps				
10mm2		Item		81.60
		Carried forward		620.82

	Quantity	Unit	Rate	Total £
		Brought forward		620.82
Supply and install cross-bonding to water services (sinks, boilers, etc.) including earth bonding clamps				
6mm2		Item		41.46
LV SWITCH GEAR AND DISTRIBUTION				
Remove existing consumer unit including the disconnection of the meter and sub-circuits and supply and install new split-load consumer unit complete with integral isolator, RCCB, MCBs; all suitable rated				
SP and N; 15 module split-load insulated consumer unit fitted with various MCBs, the remainder fitted with blanks; to backgrounds requiring fixings				
consumer unit	1	each	183.31	183.31
extra for				
6 amp, SP MCB	1	each	10.55	10.55
32 amp, SP MCB	1	each	10.11	10.11
45 amp, SP MCB	1	each	10.11	10.11
bell transformer	1	each	23.45	23.45
LUMINAIRES AND LAMPS				
Replace and renew existing luminaire including the disconnection of the existing cables and re-connection of the new luminaire				
		Carried forward		899.81

	Quantity	Unit	Rate	Total £
		Brought forward		899.81
White plastic ceiling rose, lamp holder complete with white flexible PVC insulated cable not exceeding 225mm in length; to backgrounds requiring fixings				
pendant	5	each	7.70	38.50
White plastic batten lamp holder complete with skirt; to backgrounds requiring fixings				
ceiling lamp holder	2	each	7.81	15.62
Tungsten luminaire IP44 rated complete with GLS lamp and perspex diffuser; to backgrounds requiring fixings				
60 watt chrome dome, zones 1-3	1	each	25.44	25.44
Fluorescent luminaire complete with lamp and perspex diffuser; to backgrounds requiring fixings				
1500mm long; single tube	1	each	32.78	32.78
Disconnect, remove, test and re-install tenant's own decorative metal luminaire				
tenant's luminaire (ceiling)	2	each	17.57	35.14
Install tenant's own wall-mounted luminaire; to backgrounds requiring fixings				
tungsten wall lights	2	each	11.44	22.88
		Carried forward		1,070.17

	Quantity	Unit	Rate	Total £
		Brought forward		1,070.17
ACCESSORIES FOR ELECTRICAL SERVICES				
Replace and renew existing light switch plate including the disconnection of the existing cables and re-connection of the new switch plate; to existing back boxes				
White plastic switch plates				
1 gang; 1 way; single pole	6	each	5.02	30.12
3 gang; 2 way; single pole	1	each	14.80	14.80
1 gang; 2 way; single pole	5	each	5.80	29.00
Replace existing general power accessories including the disconnection of the existing cables and re-connection of the new outlet plate and the installation of flexible PVC insulated earth continuity conductor; to existing back boxes				
White plastic outlet plates				
13 amp; single switched socket outlet	2	each	7.26	14.52
13 amp; twin switched socket outlet	8	each	9.22	73.76
20 amp; double pole indicated switch engraved 'Immersion heater'	1	each	16.39	16.39
13 amp; switch fuse connection unit including outgoing final connections	2	each	12.45	24.90
45 amp; cooker control unit	1	each	19.00	19.00
low-level cooker connection unit	1	each	12.05	12.05
45 amp; double pole; indicated pull switch	1	each	24.09	24.09
		Carried forward		1,328.80

	Quantity	Unit	Rate	Total £
		Brought forward		1,328.80
250 volt grade flush-mounted accessories including back boxes, flexible PVC insulated earth continuity conductor between box and face plate; to backgrounds requiring fixings				
Switch socket outlets				
13 amp; 1 gang	2	each	9.49	18.98
13 amp; 2 gang	8	each	13.26	106.08
Switch fuse indicated unit including outgoing final connections				
13amp; double pole	4	each	14.93	59.72
Double pole indicated switch engraved 'Boiler'				
20 amp; double pole	1	each	22.86	22.86
Unswitched fuse connection unit including outgoing final connections				
13 amp	1	each	11.08	11.08
Sundry works				
Door chimes and bell push including associated wiring and protection; to backgrounds requiring fixings				
door bell system	1	each	36.98	36.98
		Carried forward		1,584.50

	Quantity	Unit	Rate	Total £
		Brought forward		1,584.50
240/9 volt grade surface-mounted smoke detector and fire alarm unit complete with rechargeable battery for back-up purposes; hard wired; to backgrounds requiring fixings				
Smoke detectors/fire alarm unit complete with associated PVC insulated PVC sheathed cable taken from local lighting circuit				
smoke detector/fire alarm unit	2	each	33.59	67.18
240/9 volt grade surface-mounted carbon monoxide detector and fire alarm unit complete with re-chargeable battery for back-up purposes; hard wired; to backgrounds requiring fixings				
Carbon monoxide detector unit complete with associated PVC insulated PVC sheathed cable taken from local power circuit via fuse connection unit				
carbon monoxide detector unit	1	each	54.27	54.27
		Carried forward		1,705.95

	Quantity	Unit	Rate	Total £
		Brought forward		1,705.95
EXTERNAL LIGHTING				
FITTINGS AND ACCESSORIES				
250 volt grade tungsten luminaire complete with lamp; as manufactured by 'Coughtrie'				
Surface-mounted bulkhead luminaire including conduit sleeve through wall, bushing, final connections and the like; to backgrounds requiring fixings				
'Coughtrie' bulkhead luminaire	2	each	43.80	87.60
Final testing				
Final testing as per IEE Wiring Regulations and production of standard NICEIC documentation per dwelling				
Three bedroom, six person house	1	each	60.00	60.00
Operating and training				
Demonstrate to each tenant the RCCB and MCBs in operation and hand to each tenant a prepared leaflet covering this operation	1	each	6.00	6.00
		Carried forward		1,859.55

	Quantity	Unit	Rate	Total £
		Brought forward		1,859.55
BUILDER'S WORK IN CONNECTION WITH ELECTRICAL SERVICES				
Cut out in brick or block walls, and make good plaster for				
single socket outlet and/or fuse connection unit	6	each	5.35	32.10
twin socket outlet	8	each	6.50	52.00
low-level cooker connection unit and/or deep pattern socket box	1	each	5.35	5.35
Cut chase in brick or block wall, make good plaster 50mm wide				
cable chase	45	m	1.78	80.10
Floor boards, tongued and grooved				
lift softwood flooring; using hand tools	15	m	6.25	93.75
relay softwood flooring; fixing with screws	15	m	2.07	31.05
TOTAL OF UPGRADING OF ELECTRICAL SERVICES TO THREE BEDROOM, SIX PERSON HOUSE				2,153.90

	Quantity	Unit	Rate	Total £
FOUR BEDROOM, SIX PERSON HOUSE				
TESTING				
Electric tests				
100% of circuits		Item		36.00
extra for visual inspection		Item		20.75
extra for complete schedules and report		Item		20.75
REMEDIAL WORK				
LV/HV CABLES AND WIRING				
600/1000 volt grade PVC insulated, PVC sheathed cable, single core				
Remove existing meter tails and renew including liaison with local electricity authority/shipper for accessing meter and re-sealing on completion				
25mm2		Item		27.00
600/1000 volt grade PVC insulated, PVC sheathed; including PVC capping where necessary				
Drawn into voids or chases or clipped to backgrounds; twin and earth cabling				
1.5mm2	20	m	2.28	45.60
2.5mm2	50	m	2.74	137.00
6.0mm2	20	m	7.43	148.60
		Carried forward		435.70

	Quantity	Unit	Rate	Total £
		Brought forward		435.70
Drawn into voids or chases or clipped to backgrounds; three core and earth cabling				
1.5mm2	18	m	4.42	79.56
300/500 volt grade, PVC insulated PVC sheathed heat resistant circular cables; copper stranded conductors; BS 6141; in tails (not exceeding 1 metre long) including termination at both ends				
Remove existing immersion heater and boiler cables (in tails) and renew				
3 core; 2.5mm2		Item		29.60
EARTHING AND BONDING COMPONENTS				
Copper earth connection including copper insulated cables and final connection and PVC capping where necessary				
Remove main earth between incoming main and consumer unit and renew				
16mm2		Item		24.12
Supply and install main equipotential earth bonding between incoming mains and other incoming services including earth bonding clamps				
10mm2		Item		81.60
		Carried forward		650.58

	Quantity	Unit	Rate	Total £
		Brought forward		650.58
Supply and install cross-bonding to water services (sinks, boilers, etc.) including earth bonding clamps				
6mm2		Item		41.46
LV SWITCH GEAR AND DISTRIBUTION				
Remove existing consumer unit including the disconnection of the meter and sub-circuits and supply and install new split-load consumer unit complete with integral isolator, RCCB, MCBs; all suitable rated				
SP and N; 15 module split-load insulated consumer unit fitted with various MCBs, the remainder fitted with blanks; to backgrounds requiring fixings				
consumer unit	1	each	183.31	183.31
extra for				
6 amp, SP MCB	1	each	10.55	10.55
16 amp, SP MCB	1	each	10.11	10.11
32 amp, SP MCB	1	each	10.11	10.11
bell transformer	1	each	23.45	23.45
LUMINAIRES AND LAMPS				
Replace and renew existing luminaire including the disconnection of the existing cables and re-connection of the new luminaire				
		Carried forward		929.57

	Quantity	Unit	Rate	Total £
		Brought forward		929.57
White plastic ceiling rose, lamp holder complete with white flexible PVC insulated cable not exceeding 225mm in length; to backgrounds requiring fixings				
pendant	8	each	7.70	61.60
White plastic batten lamp holder complete with skirt; to backgrounds requiring fixings				
ceiling lamp holder	2	each	7.81	15.62
Tungsten luminaire IP44 rated complete with GLS lamp and perspex diffuser; to backgrounds requiring fixings				
60 watt chrome dome, zones 1-3	1	each	25.44	25.44
Fluorescent luminaire complete with lamp and perspex diffuser; to backgrounds requiring fixings				
1500mm long; single tube	1	each	32.78	32.78
ACCESSORIES FOR ELECTRICAL SERVICES				
Replace and renew existing light switch plate including the disconnection of the existing cables and re-connection of the new switch plate; to existing back boxes				
White plastic switch plates				
1 gang; 1 way; single pole	7	each	5.02	35.14
3 gang; 2 way; single pole	1	each	14.80	14.80
1 gang; 2 way; single pole	1	each	5.80	5.80
2 gang; 2 way; single pole	1	each	8.47	8.47
		Carried forward		1,129.22

	Quantity	Unit	Rate	Total £
		Brought forward		1,129.22
White plastic pull switches including base plate				
1 way; single pole; 6 amp	1	each	10.17	10.17
Replace existing general power accessories including the disconnection of the existing cables and re-connection of the new outlet plate and the installation of flexible PVC insulated earth continuity conductor; to existing back boxes				
White plastic outlet plates				
13 amp; single switched socket outlet	2	each	7.26	14.52
13 amp; twin switched socket outlet	18	each	9.22	165.96
20 amp; double pole indicated switch engraved 'Immersion heater'	1	each	16.39	16.39
13 amp; switch fuse connection unit including outgoing final connections	2	each	12.45	24.90
45 amp; cooker control unit	1	each	19.00	19.00
13 amp unswitched fuse connection unit including outgoing final connections	1	each	11.90	11.90
250 volt grade flush-mounted accessories including back boxes, flexible PVC insulated earth continuity conductor between box and face plate; to backgrounds requiring fixings				
Switch socket outlets				
13 amp; 1 gang	3	each	9.49	28.47
13 amp; 2 gang	8	each	13.26	106.08
		Carried forward		1,526.61

	Quantity	Unit	Rate	Total £
		Brought forward		1,526.61
Switch fuse indicated unit including outgoing final connections				
13 amp; double pole	2	each	14.93	29.86
Low-level cooker connection unit including final connections	1	each	13.30	13.30
Double pole indicated switch engraved 'Boiler'				
20 amp; double pole	1	each	22.86	22.86
Sundry works				
Door chimes and bell push including associated wiring and protection; to backgrounds requiring fixings				
door bell system	1	each	36.98	36.98
240/9 volt grade surface-mounted smoke detector and fire alarm unit complete with rechargeable battery for back-up purposes; hard wired; to backgrounds requiring fixings				
		Carried forward		1,629.61

	Quantity	Unit	Rate	Total £
		Brought forward		1,629.61
Smoke detectors/fire alarm unit complete with associated PVC insulated PVC sheathed cable taken from local lighting circuit				
smoke detector/fire alarm unit	2	each	33.59	67.18
EXTERNAL LIGHTING				
FITTINGS AND ACCESSORIES				
250 volt grade tungsten luminaire complete with lamp; as manufactured by 'Coughtrie'				
Surface-mounted bulkhead luminaire including conduit sleeve through wall, bushing, final connections and the like; to backgrounds requiring fixings				
'Coughtrie' bulkhead luminaire	2	each	43.80	87.60
Final testing				
Final testing as per IEE Wiring Regulations and production of standard NICEIC documentation per dwelling				
Four bedroom, six person house	1	each	60.00	60.00
		Carried forward		1,844.39

	Quantity	Unit	Rate	Total £
		Brought forward		1,844.39
Operating and training				
Demonstrate to each tenant the RCCB and MCBs in operation and hand to each tenant a prepared leaflet covering this operation	1	each	6.00	6.00
BUILDER'S WORK IN CONNECTION WITH ELECTRICAL SERVICES				
Cut out in brick or block walls, and make good plaster for				
single socket outlet and/or fuse connection unit	6	each	5.35	32.10
twin socket outlet	8	each	6.50	52.00
low-level cooker connection unit and/or deep pattern socket box	1	each	5.35	5.35
Cut chase in brick or block wall, make good plaster 50mm wide				
cable chase	20	m	1.78	35.60
Floor boards, tongued and grooved				
lift softwood flooring; using hand tools	12	m	6.25	75.00
relay softwood flooring; fixing with screws	12	m	2.07	24.84
TOTAL OF UPGRADING OF ELECTRICAL SERVICES TO FOUR BEDROOM, SIX PERSON HOUSE				2,075.28

REWIRING

The work comprised the total rewiring of Housing Association properties including the provision of additional points and fittings. Rates, particularly cables, reflect small one-off properties on various estates, therefore, a small volume of cables and accessories being used, consequently mean that only small discounts would be obtained.

The existing wiring shall be withdrawn and removed, and all visible cables are also removed. All new work shall be installed flush, either 'fished' through dry linings or covered by suitably sized heavy duty plastic capping to protect the cables.

The Contractor is responsible for all the builder's work in connection with the rewiring, i.e. moving and replacing furniture, cutting chases, making good plaster and lifting and relaying floorboards and carpeting. The Contractor carry out all the making good of plasterwork.

The properties are occupied during the whole of the works and due allowance shall be made for maintaining power to refrigerators and freezers at all times as well as lighting during the evening. The Contractor is to reasonable notice and an assessment of the length of time that the power will be 'off' for the remaining circuits. The contractor is responsible for re-setting all those items of equipment that are affected by the turning 'off' of the power, e.g. time clocks.

It has been assumed that the Contractor is a member of the National Inspection Council for Electrical Installation Contracting and carries out the work in accordance with the 16th edition of the Insitute of Electrical Engineers' Wiring Regulations.

An allowance has been for wiring to the wall above the end of the bath for a shower and for the adequate loop of wiring in the roof space for the future connection of a ceiling-mounted pull switch. A blanking plate, back box and separate circuit breaker is installed where an electric shower is not fitted.

Each property is to be fitted with a new consumer unit with split-load, integral isolator suitably rated, RCCB, MCBs, and blanks and all circuits shall be clearly labelled. Each property is fitted with smoke detectors, complete with rechargeable batteries, each hard wired to the local lighting circuit and interconnected. The detectors are ceiling mounted and located in the hall and first floor landing in the houses and in the hall in the bungalows.

The new installation shall allow for switches mounted 1200mm above finished floor level and socket outlets, except in the kitchen, 600mm above finished floor level. Outlets in kitchens are 200mm above worktop level generally. Appliances housed beneath worktops had their socket outlets fitted 600mm above finished floor level and are independently controlled by a switched fuse connection unit to line up with the outlets above worktop level.

Lighting shall be provided to all rooms as follows:

- Pendant fittings to living rooms, dining rooms, porch/hall, bedrooms and internal porches.
- Single 1500mm fluorescent fitting with diffuser to kitchens.
- Batten fittings to landings, WCs and stores.
- Zones 1-3 totally enclosed luminaire within bathrooms
- External bulkhead fitting adjacent to rear and front doors.
- Two way switches between halls and landings.

All earth bonding and cross-bonding within each property and the upgrading of the main earth shall be carried out. Mains-operated carbon monoxide detector are fitted to each property that has a gas fire gitted and this work is priced separately.

The power for the detector is derived from the local power circuit via unswitched fuse connection unit, both mounted at level/distance from the gas fire as recommended by the maunfacturers.

Each property is provided with the following,, as a minimum requirement. Other equipment or fittings that have been added to the installation by the tenant are also rewired; this work being treated as a variation to the contract.

Three bedroom, four person house; Type A

Ground floor

Living room

2 no. pendant light fittings
4 no. double socket outlets; switched
2 no. plate switches
1 no. carbon monoxide detector with fuse connection unit (where gas-fire fitted)

Kitchen

1 no. fluorescent fitting
1 no. plate switch
4 no. double socket outlets; switched
1 no. cooker control unit with connection unit
1 no. immersion switch with neon indicator
3 no. switched spurs
3 no. single socket outlets; switched

Dining room

1 no. plate switch
2 no. double socket outlets; switched
1 no. pendant light fitting

Porch/hall
1 no. pendant light fitting
1 no. plate switch (hall, landing and outside light, if fitted)
1 no. double socket outlet; switched
1 no. smoke alarm (interconnected)
1 no. bulkhead fitting; external (if fitted)

Store/boiler cupboard
1 no. switched spur for boiler

Rear vestibule
1 no. pendant light fitting
1 no. plate switch (vestibule and outside light)
1 no. bulkhead fitting; external

First floor

Landing
1 no. pendant light fitting
1 no. plate switch
1 no. smoke alarm (interconnected)
1 no. single socket outlet; switched

Bedroom 1
2 no. double socket outlets, switched
1 no. pendant light fitting
1 no. plate switch

Bedroom 2
2 no. double socket outlets; switched
1 no. pendant light fitting
1 no. plate switch

Bedroom 3
1 no. double socket outlet; switched
1 no. pendant light fitting
1 no. plate switch

Bathroom
1 no. pull switch
1 no. tungsten luminaire; Zones 1-3 inclusive
1 no. pull switch for shower

Cylinder cupboard
- 1 no. double pole switched connection unit with flex outlet for immersion heater

Garage
- 2 no. batten light fitting
- 1 no. surface-mounted plate switch
- 1 no. surface-mounted twin socket outlet; switched
- 1 no. surface-mounted fuse connection unit; switched

Three bedroom, five person house; Type B

Ground floor

Living room
- 2 no. pendant light fittings
- 4 no. double socket outlets; switched
- 2 no. plate switches
- 1 no. carbon monoxide detector unswitched fuse connection unit (where gas-fire fitted)

Kitchen
- 1 no. fluorescent fitting
- 1 no. plate switch
- 4 no. double socket outlets; switched
- 1 no. cooker control unit with connection unit
- 1 no. immersion switch with neon indicator
- 3 no. switched spurs
- 3 no. single socket outlets; switched

Dining room
- 1 no. plate switch
- 1 no. single socket outlet; switched
- 2 no. double socket outlets; switched
- 1 no. pendant light fitting

Hall
- 1 no. pendant light fitting
- 1 no. plate switch (hall, landing and outside light)
- 1 no. double socket outlet; switched
- 1 no. smoke alarm (interconnected)
- 1 no. bulkhead fitting; external

Store/boiler cupboard
- 1 no. switched spur for boiler

WC
1 no. batten lamp holder
1 no. plate switch (external to WC)

Rear vestibule
1 no. pendant light fitting
1 no. plate switch (vestibule and outside light)
1 no. bulkhead fitting; external

First floor

Landing
1 no. pendant light fitting
1 no. plate switch
1 no. smoke alarm (interconnected)
1 no. single socket outlet; switched

Bedroom 1
2 no. double socket outlets; switched
1 no. pendant light fitting
1 no. plate switch

Bedroom 2
2 no. double socket outlets; switched
1 no. pendant light fitting
1 no. plate switch

Bedroom 3
1 no. single socket outlet; switched
1 no. double socket outlet; switched
1 no. pendant light fitting
1 no. plate switch

Bathroom
1 no. pull switch
1 no. tungsten luminaire; Zones 1-3 inclusive
1 no. pull switch for shower

WC
1 no. batten lamp holder
1 no. plate switch (external to WC)

Cylinder cupboard

1 no. double pole switched connection unit with flex outlet for immersion heater

Garage

1 no. batten light fitting
2 no. surface-mounted plate switch
1 no. surface-mounted twin socket outlet; switched
1 no. surface-mounted fuse connection unit; switched

Four bedroom, seven person house; Type C

Ground floor

Living room

2 no. pendant light fittings
4 no. double socket outlets; switched
2 no. plate switches
1 no. carbon monoxide detector unswitched fuse connection unit (where gas-fire fitted)

Kitchen

1 no. fluorescent fitting
1 no. plate switch
4 no. double socket outlets; switched
1 no. cooker control unit with connection unit
1 no. immersion switch with neon indicator
3 no. switched spurs
3 no. single socket outlets; switched

Dining room

1 no. plate switch
1 no. single socket outlet; switched
2 no. double socket outlets; switched
1 no. pendant light fitting

Hall

1 no. pendant light fitting
1 no. plate switch (hall, landing and outside light)
1 no. double socket outlet; switched
1 no. smoke alarm (interconnected)
1 no. bulkhead fitting; external

Store/boiler cupboard
1 no. switched spur for boiler

WC
1 no. batten lamp holder
1 no. plate switch (external to WC)

Rear vestibule
1 no. pendant light fitting
1 no. plate switch (vestibule and external light)
1 no. bulkhead fitting; external
1 no. single socket outlet; switched

First floor
Landing
1 no. pendant light fitting
1 no. plate switch
1 no. smoke alarm (interconnected)
1 no. single socket outlet; switched

Bedroom 1
2 no. double socket outlets; switched
1 no. pendant light fitting
1 no. plate switch

Bedroom 2
2 no. double socket outlets; switched
1 no. pendant light fitting
1 no. plate switch

Bathroom
1 no. pull switch
3 no. tungsten luminaire; Zones 1-3 inclusive
1 no. pull switch for shower

WC
1 no. batten lamp holder
1 no. plate switch (external to WC)

Cylinder cupboard
1 no. double pole switched connection unit with flex outlet for immersion heater

Second floor

Landing

1 no. pendant light fitting
1 no. plate switch
1 no. smoke alarm (interconnected)
1 no. single socket outlet; switched

Bedroom 3

2 no. double socket outlet; switched
1 no. pendant light fitting
1 no. plate switch

Bedroom 4

1 no. single socket outlet; switched
1 no. double socket outlet; switched
1 no. pendant light fitting
1 no. plate switch

Bathroom

1 no. pull switch
1 no. tungsten luminaire; Zones 1-3 inclusive

One bedroom, two person bungalow; Type D

Ground floor

Hall

1 no. pendant light fitting
1 no. plate switch
1 no. double socket outlet; switched
1 no. smoke alarm (interconnected)

Living room

3 no. double socket outlets; switched
2 no. pendant light fittings
2 no. plate switches

Kitchen

1 no. batten lamp holder
1 no. plate switch
2 no. double socket outlets; switched

Kitchen (cont'd)

1 no. cooker control unit with connection unit
1 no. immersion switch with neon indicator
2 no. switched spurs
2 no. single socket outlets; switched
1 no. isolating switch for extract fan

Store/boiler cupboard
1 no. switched spur for boiler

Rear vestibule
1 no. pendant light fitting
1 no. plate switch (vestibule and outside light)
1 no. bulkhead fitting; external

Bathroom
1 no. pull switch
1 no. tungsten luminaire; Zones 1-3 inclusive
1 no. pull switch for shower
1 no. isolating switch for extract fan

Bedroom
1 no. pendant light fitting
1 no. plate switch
2 no. double socket outlets; switched

Cylinder cupboard
1 no. double pole switched connection unit with flex outlet for immersion heater

	Quantity	Unit	Rate	Total £
REWIRING OF THREE BEDROOM, FOUR PERSON HOUSE				
EXISTING SERVICES				
Isolate, disconnect, cut back and remove existing wiring, accessories, fittings and the like and remove from site		Item		24.83
LV/HV CABLES AND WIRING				
600/1000 volt grade PVC insulated, PVC sheathed cable, single core				
New meter tails including liaison with local electricity authority/shipper for accessing meter and re-sealing on completion				
25mm2		Item		17.55
600/1000 volt grade PVC insulated, PVC sheathed; including PVC capping where necessary				
Drawn into voids or chases or clipped to backgrounds; twin and earth cabling				
1.5mm2	200	m	2.28	456.00
2.5mm2	150	m	2.74	411.00
6.0mm2	50	m	7.43	371.50

Carried forward 1,280.88

	Quantity	Unit	Rate	Total £
		Brought forward		1,280.88
Drawn into voids or chases or clipped to backgrounds; three core and earth cabling				
1.5mm2	15	m	5.03	75.45
300/500 volt grade, PVC insulated PVC sheathed heat resistant circular cables; copper stranded conductors; BS 6141; in tails including termination at both ends				
Supply to immersion heater and boiler				
3 core; 2.5mm2	2	m	11.69	23.38
EARTHING AND BONDING COMPONENTS				
Copper earth connection including copper insulated cables and final connection and PVC capping where necessary				
Main earth between incoming main and consumer unit				
16mm2		Item		12.53
Supply and install main equipotential earth bonding between incoming mains and other incoming services including earth bonding clamps				
10mm2		Item		81.60
		Carried forward		1,473.84

	Quantity	Unit	Rate	Total £
		Brought forward		1,473.84
Supply and install cross-bonding to water services (sinks, boilers, etc.) including earth bonding clamps				
6mm2		Item		41.46
LV SWITCH GEAR AND DISTRIBUTION				
Consumer unit; new split-load consumer unit complete with integral isolator, RCCB, MCBs; all suitable rated				
SP and N; 15 module split-load insulated consumer unit fitted with various MCBs, the remainder fitted with blanks; to backgrounds requiring fixings				
consumer unit	1	each	173.06	173.06
extra for				
6 amp, SP MCB (extra lighting and bell transformer)	2	each	10.55	21.10
16 amp, SP MCB (boiler)	1	each	10.11	10.11
32 amp, SP MCB (extra ring mains)	1	each	10.11	10.11
45 amp, SP MCB (shower)	1	each	10.11	10.11
bell transformer	1	each	23.45	23.45
LUMINAIRES AND LAMPS				
Surface-mounted internal luminaires				
White plastic ceiling rose, lamp holder complete with white flexible PVC insulated cable not exceeding 225mm in length; to backgrounds requiring fixings				
pendant	9	each	7.70	69.30
		Carried forward		1,832.54

	Quantity	Unit	Rate	Total £
		Brought forward		1,832.54
White plastic batten lamp holder complete with skirt; to backgrounds requiring fixings				
ceiling lamp holder	2	each	7.81	15.62
Tungsten luminaire; IP44 rated complete with GLS lamp and diffuser; to backgrounds requiring fixings				
60 watt chrome dome luminaire; zones 1-3	1	each	25.44	25.44
Fluorescent luminaire complete with lamp and perspex diffuser; to backgrounds requiring fixings				
1500mm long; single tube	1	each	29.4	29.40
Disconnect, remove, test and re-install tenant's own luminaire				
tenant's own luminaire (wall mounted)	2	each	17.57	35.14
ACCESSORIES FOR ELECTRICAL SERVICES				
250 volt grade flush-mounted accessories; fixed to existing flush-mounted switch boxes				
White plastic switch plates				
1 gang; 1 way; single pole	7	each	5.02	35.14
2 gang; 2 way; single pole	1	each	8.47	8.47
3 gang; 2 way; single pole	1	each	14.80	14.80
1 gang; 2 way; single pole	2	each	5.80	11.60
		Carried forward		2,008.15

	Quantity	Unit	Rate	Total £
		Brought forward		2,008.15
White plastic pull switches including base plate				
1 way; single pole	1	each	10.17	10.17
250 volt grade flush-mounted accessories; fixed to existing flush-mounted socket boxes and the installation of flexible PVC insulated earth continuity conductor between box and face plate				
White plastic outlet plates				
13 amp; single switched socket outlet	1	each	7.26	7.26
13 amp; twin switched socket outlet	7	each	9.22	64.54
20 amp; double pole indicated switch engraved 'Immersion heater'	1	each	16.37	16.37
13 amp; switch fuse connection unit including outgoing final connections	1	each	12.45	12.45
50 amp; cooker control unit	1	each	19.00	19.00
250 volt grade flush-mounted accessories including dry lining back boxes; fit into apertures				
Switch socket outlets				
13 amp; 1 gang	3	each	7.75	23.25
13 amp; 2 gang	9	each	10.78	97.02
Switch fuse indicator unit including outgoing final connections				
13 amp; double pole	3	each	13.19	39.57
		Carried forward		2,297.78

	Quantity	Unit	Rate	Total £
		Brought forward		2,297.78
Low level cooker connection unit including final connections				
connection unit	1	each	11.25	11.25
Double pole indicated switch engraved 'Immersion heater'				
20 amp; double pole	1	each	14.38	14.38
Unswitched fuse connection unit including outgoing final connections				
13 amp	1	each	16.05	16.05
250 volt grade surface-mounted accessories including moulded plastic back boxes; to backgrounds requiring fixings (garage)				
White plastic switch plate				
1 gang; 1 way; single pole	1	each	7.39	7.39
White plastic switch socket outlets				
13 amp; 2 gang	1	each	12.61	12.61
White plastic fuse connection unit; switched				
13 amp; double pole	1	each	15.01	15.01
250 volt grade surface-mounted accessories including back boxes; to backgrounds requiring fixings				
		Carried forward		2,374.47

	Quantity	Unit	Rate	Total £
		Brought forward		2,374.47
Double pole indicated pull switch				
45 amp; double pole	1	each	23.23	23.23
Sundry work				
Door chimes and bell push including associated wiring and protection; to backgrounds requiring fixings				
door bell system	1	each	36.98	36.98
250 volt/9 volt grade surface-mounted smoke detector and fire alarm unit complete with rechargeable battery for back-up purposes; hard wired; to backgrounds requiring fixings				
Smoke detector/fire alarm unit complete with associated PVC insulated PVC sheathed cable taken from local lighting circuit				
smoke detector/fire alarm unit	2	each	33.59	67.18
250 volt/9 volt grade surface-mounted carbon monoxide detector unit complete with rechargeable battery for back-up purposes; hard wired; to backgrounds requiring fixings				
Carbon monoxide detector unit complete with associated PVC insulated PVC sheathed cable taken from local power circuit via fuse connection unit				
		Carried forward		2,501.86

	Quantity	Unit	Rate	Total £
		Brought forward		2,501.86
carbon monoxide detector unit	1	each	54.17	54.17
EXTERNAL LIGHTING				
FITTINGS AND ACCESSORIES				
250 volt grade tungsten luminaire complete with lamp; as manufactured by 'Coughtrie)				
Surface-mounted bulkhead luminaire including conduit sleeve through wall, bushing; final connections and the like; to backgrounds requiring fixings				
'Coughtrie' bulkhead luminaire	1	each	43.80	43.80
Testing				
Final testing as per IEE Wiring Regulations and production of standard NICEIC documentation per dwelling				
three bedroom, four person house		Item		48.00
Operating and training				
Demonstrate to each tenant the RCCB and MCBs in operation and hand to each tenant a leaflet covering this operation		Item		6.00
		Carried forward		2,653.83

	Quantity	Unit	Rate	Total £
		Brought forward		2,653.83
BUILDER'S WORK IN CONNECTION WITH ELECTRICAL SERVICES				
Cut out in dry lining walls, and make good plaster for				
single socket outlet and/or fuse connection unit	7	each	4.16	29.12
twin socket outlet	8	each	4.16	33.28
low-level cooker connection unit and/or deep pattern socket box	2	each	4.16	8.32
Cut chase in brick or block wall, make good plaster 50mm wide				
cable capping	10	m	1.78	17.80
Floor boards; tongued and grooved				
lift softwood flooring; using hand tools	15	m	6.25	93.75
relay softwood flooring; fixing with screws	15	m	2.07	31.05
UPGRADING OF ELECTRICAL SERVICES TO THREE BEDROOM, FOUR PERSON HOUSE				2,867.15

	Quantity	Unit	Rate	Total £
REWIRING OF THREE BEDROOM FIVE PERSON HOUSE				
EXISTING SERVICES				
Isolate, disconnect, cut back and remove existing wiring, accessories, fittings and the like and remove from site		Item		24.83
LV/HV CABLES AND WIRING				
600/1000 volt grade PVC insulated, PVC sheathed cable, single core				
New meter tails including liaison with local electricity authority/shipper for accessing meter and re-sealing on completion				
25mm2		Item		17.55
600/1000 volt grade PVC insulated, PVC sheathed; including PVC capping where necessary				
Drawn into voids or chases or clipped to backgrounds; twin and earth cabling				
1.5mm2	220	m	2.28	501.60
2.5mm2	175	m	2.74	479.50
6.0mm2	50	m	7.43	371.50
		Carried forward		1,394.98

	Quantity	Unit	Rate	Total £
		Brought forward		1,394.98
Drawn into voids or chases or clipped to backgrounds; three core and earth cabling				
1.5mm2	15	m	5.03	75.45
300/500 volt grade, PVC insulated PVC sheathed heat resistant circular cables; copper stranded conductors; BS 6141; in tails including termination at both ends				
Supply to immersion heater and boiler				
3 core; 2.5mm2	2	m	11.69	23.38
EARTHING AND BONDING COMPONENTS				
Copper earth connection including copper insulated cables and final connection and PVC capping where necessary				
Main earth between incoming main and consumer unit				
16mm2		Item		12.53
Supply and install main equipotential earth bonding between incoming mains and other incoming services including earth bonding clamps				
10mm2		Item		81.60
		Carried forward		1,587.94

	Quantity	Unit	Rate	Total £
		Brought forward		1,587.94
Supply and install cross-bonding to water services (sinks, boilers, etc.) including earth bonding clamps				
6mm2		Item		41.46
LV SWITCH GEAR AND DISTRIBUTION				
Consumer unit; new split-load consumer unit complete with integral isolator, RCCB, MCBs; all suitable rated				
SP and N; 15 module split-load insulated consumer unit fitted with various MCBs, the remainder fitted with blanks; to backgrounds requiring fixings				
consumer unit	1	each	173.06	173.06
extra for				
6 amp, SP MCB (extra lighting)	1	each	10.55	10.55
16 amp, SP MCB (boiler)	1	each	10.11	10.11
32 amp, SP MCB (extra ring mains)	2	each	10.11	20.22
45 amp, SP MCB (shower)	1	each	10.55	10.55
LUMINAIRES AND LAMPS				
Surface-mounted internal luminaires				
White plastic ceiling rose, lamp holder complete with white flexible PVC insulated cable not exceeding 225mm in length; to backgrounds requiring fixings				
pendant	9	each	7.70	69.30
		Carried forward		1,923.19

	Quantity	Unit	Rate	Total £
		Brought forward		1,923.19
White plastic batten lamp holder complete with skirt; to backgrounds requiring fixings				
ceiling lamp holder	3	each	7.81	23.43
Recessed mains voltage luminaire; IP44 rated complete with opal lens and lamp; fitted into recess formed by others				
60 watt tungsten downlighter; zones 1-3	3	each	18.56	55.68
Fluorescent luminaire complete with lamp and perspex diffuser; to backgrounds requiring fixings				
1500mm long; single tube	1	each	29.40	29.40
Install tenant's own wall-mounted luminaire; to backgrounds requiring fixings				
tungsten wall lights	2	each	11.44	22.88
ACCESSORIES FOR ELECTRICAL SERVICES				
250 volt grade flush-mounted accessories; fixed to existing flush-mounted switch boxes				
White plastic switch plates				
1 gang; 1 way; single pole	7	each	5.02	35.14
2 gang; 2 way; single pole	1	each	8.47	8.47
3 gang; 2 way; single pole	1	each	14.80	14.80
1 gang; 2 way; single pole	2	each	5.80	11.60
		Carried forward		2,124.59

	Quantity	Unit	Rate	Total £
		Brought forward		2,124.59
White plastic pull switches including base plate				
1 way; single pole	1	each	10.17	10.17
250 volt grade flush-mounted accessories; fixed to existing flush-mounted switch boxes and the installation of flexible PVC insulated earth continuity conductor between box and face plate				
White plastic outlet plates				
13 amp; single switched socket outlet	2	each	7.26	14.52
13 amp; twin switched socket outlet	8	each	9.22	73.76
20 amp; double pole indicated switch engraved 'Immersion heater'	1	each	16.37	16.37
13 amp; switch fuse connection unit including outgoing final connections	1	each	12.45	12.45
50 amp; cooker control unit	1	each	19.00	19.00
250 volt grade flush-mounted accessories including dry lining back boxes; fit into apertures				
Switch socket outlets				
13 amp; 1 gang	4	each	7.75	31.00
13 amp; 2 gang	8	each	10.78	86.24
Switch fuse indicator unit including outgoing final connections				
13 amp; double pole	3	each	13.19	39.57
		Carried forward		2,427.67

	Quantity	Unit	Rate	Total £
		Brought forward		2,427.67
Low-level cooker connection unit including final connections				
connection unit	1	each	11.25	11.25
Double pole indicated switch engraved 'Immersion heater'				
20 amp; double pole	1	each	14.38	14.38
Unswitched fuse connection unit including outgoing final connections				
13 amp	1	each	16.05	16.05
250 volt grade surface-mounted accessories including moulded plastic back boxes; to backgrounds requiring fixings (garage)				
White plastic switch plate				
1 gang; 2 way; single pole	2	each	7.78	15.56
White plastic switch socket outlets				
13 amp; 2 gang	1	each	12.61	12.61
White plastic fuse connection unit; switched				
13 amp; double pole	1	each	15.01	15.01
250 volt grade surface mounted accessories including back boxes; to backgrounds requiring fixings				
Double pole indicated pull switch				
		Carried forward		2,512.53

	Quantity	Unit	Rate	Total £
		Brought forward		2,512.53
45 amp; double pole	1	each	23.23	23.23
250 volt/9 volt grade surface-mounted smoke detector and fire alarm unit complete with rechargeable battery for back-up purposes; hard wired; to backgrounds requiring fixings				
Smoke detector/fire alarm unit complete with associated PVC insulated PVC sheathed cable taken from local lighting circuit				
smoke detector/fire alarm unit	2	each	33.59	67.18
250 volt/9 volt grade surface-mounted carbon monoxide detector unit complete with rechargeable battery for back-up purposes; hard wired; to backgrounds requiring fixings				
Carbon monoxide detector unit complete with associated PVC insulated PVC sheathed cable taken from local power circuit via fuse connection unit				
carbon monoxide detector unit	1	each	54.17	54.17
EXTERNAL LIGHTING FITTINGS AND ACCESSORIES				
250 volt grade tungsten luminaire complete with lamp; as manufactured by 'Coughtrie'				
Surface-mounted bulkhead luminaire including sleeve through wall, bushing, final connections and the like; to backgrounds requiring fixings				
'Coughtrie' bulkhead luminaire	2	each	43.80	87.60
		Carried forward		2,744.71

	Quantity	Unit	Rate	Total £
		Brought forward		2,744.71
Testing				
Final testing as per IEE Wiring Regulations and production production of standard NICEIC documentation per dwelling				
three bedroom, four person house		Item		60.00
Operating and training				
Demonstrate to each tenant the RCCB and MCBs in operation and hand to each tenant a leaflet covering this operation		Item		6.00
BUILDER'S WORK IN CONNECTION WITH ELECTRICAL SERVICES				
Cut out in dry lining walls, make good plaster for				
single socket outlet and/or fuse connection unit	7	each	4.16	29.12
twin socket outlet	7	each	4.16	29.12
low-level cooker connection unit and/or deep pattern socket box	2	each	4.16	8.32
Cut chase in brick or block wall, make good plaster 50mm wide				
cable chase	12	m	1.78	21.36
Floor boards; butt joints				
lift softwood flooring; using hand tools	15	m	3.86	57.90
relay softwood flooring; fixing with screws	15	m	2.07	31.05
UPGRADING OF ELECTRICAL SERVICES TO THREE BEDROOM, FIVE PERSON HOUSE				2,987.58

	Quantity	Unit	Rate	Total £
REWIRING OF FOUR BEDROOM, SEVEN PERSON HOUSE				
EXISTING SERVICES				
Isolate, disconnect, cut back and remove existing wiring, accessories, fittings and the like and remove from site		Item		28.45
LV/HV CABLES AND WIRING				
600/1000 volt grade PVC insulated, PVC sheathed cable, single core				
New meter tails including liaison with local electricity authority/shipper for accessing meter and re-sealing on completion				
25mm2		Item		17.55
600/1000 volt grade PVC insulated, PVC sheathed; including PVC capping where necessary				
Drawn into voids or chases or clipped to backgrounds; twin and earth cabling				
1.5mm2	250	m	2.28	570.00
2.5mm2	200	m	2.74	548.00
6.0mm2	50	m	7.43	371.50
			Carried forward	1,535.50

	Quantity	Unit	Rate	Total £
		Brought forward		1,535.50
Drawn into voids or chases or clipped to backgrounds; three core and earth cabling				
1.5mm2	35	m	5.03	176.05
300/500 volt grade, PVC insulated, PVC sheathed heat resistant circular cables; copper stranded conductors; BS 6141; in tails including termination at both ends				
Supply to immersion heater and boiler				
3 core; 2.5mm2	2	m	11.69	23.38
EARTHING AND BONDING COMPONENTS				
Copper earth connection including copper insulated cables and final connection and PVC capping where necessary				
Main earth between incoming main and consumer unit				
16mm2		Item		12.53
Supply and install main equipotential earth bonding between incoming mains and other incoming services including earth bonding clamps				
10mm2		Item		81.60
		Carried forward		1,829.06

	Quantity	Unit	Rate	Total £
		Brought forward		1,829.06
Supply and install cross-bonding to water services (sinks, boilers, etc.) including earth bonding clamps				
6mm2		Item		58.74
LV SWITCH GEAR AND DISTRIBUTION				
Consumer unit; new split-load consumer unit complete with integral isolator, RCCB, MCBs; all suitable rated				
SP and N; 17 module split-load insulated consumer unit fitted with various MCBs, the remainder fitted with blanks; to backgrounds requiring fixings				
consumer unit	1	each	185.26	185.26
extra for				
6 amp, SP MCB (extra lighting and bell transformer)	3	each	10.55	31.65
16 amp, SP MCB (boiler)	1	each	10.11	10.11
32 amp, SP MCB (extra ring mains)	2	each	10.11	20.22
45 amp, SP MCB (shower)	1	each	10.55	10.55
bell transformer	1	each	23.45	23.45
LUMINAIRES AND LAMPS				
Surface-mounted internal luminaires				
White plastic ceiling rose, lamp holder complete with white flexible PVC insulated cable not exceeding 225mm in length; to backgrounds requiring fixings				
pendant	11	each	7.70	84.70
		Carried forward		2,253.74

	Quantity	Unit	Rate	Total £
		Brought forward		2,253.74
White plastic batten lamp holder complete with skirt; to backgrounds requiring fixings				
ceiling lamp holder	2	each	7.81	15.62
Tungsten luminaire; IP44 rated complete with GLS lamp and diffuser; to backgrounds requiring fixings				
60 watt chrome dome luminaire; zones 1-3	2	each	25.44	50.88
Fluorescent luminaire complete with lamp and perspex diffuser; to backgrounds requiring fixings				
1500mm long; single tube	1	each	29.40	29.40
ACCESSORIES FOR ELECTRICAL SERVICES				
250 volt grade flush-mounted accessories; fixed to existing flush-mounted switch boxes				
White plastic switch plates				
1 gang; 1 way; single pole	9	each	5.02	45.18
2 gang; 2 way; single pole	1	each	8.47	8.47
3 gang; 2 way; single pole	1	each	14.80	14.80
1 gang; 2 way; single pole	2	each	5.80	11.60
1 gang; intermediate; single pole	1	each	11.51	11.51
White plastic pull switches including base plate				
1 way; single pole	2	each	10.17	20.34
		Carried forward		2,461.54

	Quantity	Unit	Rate	Total £
		Brought forward		2,461.54
250 volt grade flush-mounted accessories; fixed to existing flush-mounted socket boxes and the installation of flexible PVC insulated earth conductor between box and face plate				
White plastic outlet plates				
13 amp; single switched socket outlet	2	each	7.26	14.52
13 amp; twin switched socket outlet	9	each	9.22	82.98
20 amp; double pole indicated switch engraved 'Immersion heater'	1	each	16.37	16.37
13 amp; switch fuse connection unit including outgoing final connections	1	each	12.45	12.45
50 amp; cooker control unit	1	each	19.00	19.00
low-level cooker connection unit	1	each	12.05	12.05
250 volt grade flush-mounted accessories including back boxes; flexible PVC insulated earth continuity conductor between box and face plate; to backgrounds				
Switch socket outlets				
13 amp; 1 gang	6	each	9.12	54.72
13 amp; 2 gang	9	each	12.68	114.12
Switch fuse indicator unit including outgoing final connections				
13 amp; double pole	3	each	20.05	60.15
Double pole indicated switch engraved 'Immersion heater'				
20 amp; double pole	1	each	21.60	21.60
Unswitched fuse connection unit including outgoing final connections				
		Carried forward		2,869.50

	Quantity	Unit	Rate	Total £
		Brought forward		2,869.50
13 amp	1	each	16.90	16.90
250 volt grade surface-mounted accessories including back boxes; to backgrounds requiring fixings				
Double pole indicator pull switch				
45 amp; double pole	1	each	23.23	23.23
Sundry work				
Door chimes and bell push including associated wiring and protection; to backgrounds requiring fixings				
door bell system	1	each	36.98	36.98
250/9 volt grade surface-mounted smoke detector and fire alarm unit complete with rechargeable battery for back-up purposes; hard wired; to backgrounds requiring fixings				
Smoke detector/fire alarm unit complete with associated PVC insulated PVC sheathed cable taken from local lighting circuit				
smoke detector/fire alarm unit	3	each	33.59	100.77
250/9 volt grade surface-mounted carbon monoxide detector unit complete with rechargeable battery for back-up purposes; hard wired; to backgrounds requiring fixings				
		Carried forward		3,030.48

	Quantity	Unit	Rate	Total £
		Brought forward		3,030.48
Carbon monoxide detector unit complete with associated PVC insulated PVC sheathed cable taken from local power circuit via fuse connection unit				
carbon monoxide detector unit	1	each	54.17	54.17
EXTERNAL LIGHTING				
FITTINGS AND ACCESSORIES				
250 volt grade tungsten luminaire complete with lamp; as manufactured by 'Coughtrie'				
Surface-mounted bulkhead luminaire including sleeve through wall, bushing, final connections and the like; to backgrounds requiring fixings				
'Coughtrie' bulkhead luminaire	2	each	43.80	87.60
Testing				
Final testing as per IEE Wiring Regulations and production of standard NICEIC documentation per dwelling				
three bedroom, four person house		Item		60.00
Operating and training				
Demonstrate to each tenant the RCCB and MCBs in operation and hand to each tenant a leaflet covering this operation		Item		6.00
		Carried forward		3,238.25

	Quantity	Unit	Rate	Total £
		Brought forward		3,238.25
BUILDER'S WORK IN CONNECTION WITH ELECTRICAL SERVICES				
Cut out in brick or block walls, and make good plaster for				
single socket outlet and/or fuse connection unit	9	each	5.35	48.15
twin socket outlet	8	each	6.50	52.00
low-level cooker connection unit and/or deep pattern socket box	1	each	5.35	5.35
Cut chase in brick or block walls, make good plaster 50mm wide				
cable chase	60	m	1.78	106.80
Floor boards; tongued and grooved				
lift softwood flooring; using hand tools	25	m	6.25	156.25
relay softwood flooring; fixing with screws	25	m	2.07	51.75
UPGRADING OF ELECTRICAL SERVICES TO FOUR BEDROOM, SEVEN PERSON HOUSE				3,658.55

	Quantity	Unit	Rate	Total £
REWIRING OF ONE BEDROOM, TWO PERSON BUNGALOW				
EXISTING SERVICES				
Isolate, disconnect, cut back and remove existing wiring, accessories, fittings and the like and remove from site		Item		19.00
LV/HV CABLES AND WIRING				
600/1000 volt grade PVC insulated, PVC sheathed cable, single core				
New meter tails including liaison with local electricity authority/shipper for accessing meter and re-sealing on completion				
25mm2		Item		17.55
600/1000 volt grade PVC insulated, PVC sheathed; including PVC capping where necessary				
Drawn into voids or chases or clipped to backgrounds; twin and earth cabling				
1.5mm2	100	m	2.28	228.00
2.5mm2	50	m	2.74	137.00
6.0mm2	30	m	7.43	222.90
		Carried forward		624.45

	Quantity	Unit	Rate	Total £
		Brought forward		624.45
Drawn into voids or chases or clipped to backgrounds; three core and earth cabling				
1.5mm2	15	m	5.03	75.45
300/500 volt grade, PVC insulated, PVC sheathed heat resistant circular cables; copper stranded conductors; BS 6141; in tails including termination at both ends				
Supply to immersion heater and boiler				
3 core; 2.5mm2	2	m	11.69	23.38
EARTHING AND BONDING COMPONENTS				
Copper earth connection including copper insulated cables and final connection and PVC capping where necessary				
Main earth between incoming main and consumer unit				
16mm2		Item		12.53
Supply and install main equipotential earth bonding between incoming mains and other incoming services including earth bonding clamps				
10mm2		Item		56.30
		Carried forward		792.11

	Quantity	Unit	Rate	Total £
		Brought forward		792.11
Supply and install cross-bonding to water services (sinks, boilers, etc.) including earth bonding clamps				
6mm2		Item		27.16
LV SWITCH GEAR AND DISTRIBUTION				
Consumer unit; new split-load consumer unit complete with integral isolator, RCCB, MCBs; all suitable rated				
SP and N; 12 module split-load insulated consumer unit fitted with various MCBs, the remainder fitted with blanks; to backgrounds requiring fixings				
consumer unit	1	each	164.77	164.77
extra for				
6 amp, SP MCB 9 (bell transformer)	1	each	10.55	10.55
16 amp, SP MCB (boiler)	1	each	10.11	10.11
45 amp, SP MCB (shower)	1	each	10.55	10.55
bell transformer	1	each	23.47	23.47
LUMINAIRES AND LAMPS				
Surface-mounted internal luminaires				
White plastic ceiling rose, lamp holder complete with white flexible PVC insulated cable not exceeding 225mm in length; to backgrounds requiring fixings				
pendant	5	each	7.7	38.50
		Carried forward		1,077.22

	Quantity	Unit	Rate	Total £
		Brought forward		1,077.22
White plastic batten lamp holder complete with skirt; to backgrounds requiring fixings				
ceiling lamp holder	1	each	7.81	7.81
Tungsten luminaire; IP44 rated complete with GLS lamp and diffuser; to backgrounds requiring fixings				
60 watt chrome dome luminaire; zones 1-3	1	each	25.44	25.44
ACCESSORIES FOR ELECTRICAL SERVICES				
250 volt grade flush-mounted accessories; fixed to existing flush-mounted switch boxes				
White plastic switch plates				
1 gang; 1 way; single pole	3	each	5.02	15.06
2 gang; 2 way; single pole	1	each	8.47	8.47
1 gang; 2 way; single pole	2	each	5.80	11.60
White plastic pull switches including base plate				
1 way; single pole	1	each	10.17	10.17
250 volt grade flush-mounted accessories; fixed to existing flush-mounted socket boxes				
		Carried forward		1,155.77

	Quantity	Unit	Rate	Total £
		Brought forward		1,155.77
White plastic outlet plates				
13 amp; twin switched socket outlet	3	each	9.22	27.66
20 amp; double pole indicated switch engraved 'Immersion heater'	1	each	16.37	16.37
13 amp; switch fuse connection unit including outgoing final connections	1	each	12.45	12.45
50 amp; cooker control unit	1	each	19.00	19.00
low-level cooker connection unit	1	each	12.05	12.05
250 volt grade flush-mounted accessories including back boxes; flexible PVC insulated earth continuity conductor between box and face plate; to backgrounds requiring fixings				
Switch socket outlets				
13 amp; 1 gang	2	each	9.12	18.24
13 amp; 2 gang	5	each	12.68	63.40
Switch fuse indicator unit including outgoing final connections				
13 amp; double pole	2	each	13.19	26.38
Double pole indicated switch engraved 'Immersion heater'				
20 amp; double pole	1	each	14.38	14.38
Unswitched fuse connection unit including outgoing final connections				
13 amp	1	each	16.05	16.05
		Carried forward		1,381.75

	Quantity	Unit	Rate	Total £
		Brought forward		1,381.73
White plastic fan isolator switch				
6 amp; three pole isolator switch engraved 'Fan'	1	each	21.76	21.76
250 volt grade surface-mounted accessories including back boxes; to backgrounds requiring fixings				
Double pole indicator pull switch				
45 amp; double pole	1	each	23.23	23.23
Sundry works				
Door chimes and bell push including associated wiring and protection; to backgrounds requiring fixings				
door bell system	1	each	37.95	37.95
240/9 volt grade surface-mounted smoke detector and fire alarm unit complete with rechargeable battery for back-up purposes; to backgrounds requiring fixings				
Smoke detector/fire alarm unit complete with associated PVC insulated, PVC sheathed cable taken from local lighting circuit				
smoke detector/fire alarm unit	1	each	33.59	33.59
240/9 volt grade surface-mounted carbon monoxide detector unit complete with rechargeable battery for back-up purposes; to backgrounds requiring fixings				
		Carried forward		1,498.26

	Quantity	Unit	Rate	Total £
		Brought forward		1,498.26
Carbon monoxide detector unit complete with associated PVC insulated, PVC sheathed cable taken from local power circuit via fuse connection unit				
carbon monoxide detector unit	1	each	54.17	54.17
EXTERNAL LIGHTING				
FITTINGS AND ACCESSORIES				
250 volt grade tungsten luminaire complete with lamp; as manufactured by 'Coughtrie'				
Surface-mounted bulkhead luminaire including sleeve through wall, bushing, final connections and the like; to backgrounds requiring fixings				
'Coughtrie' bulkhead luminaire	1	each	43.80	43.80
Testing				
Final testing as per IEE Wiring Regulations and production of standard NICEIC documentation per dwelling				
one bedroom, two person bungalow		Item		42.00
Operating and training				
Demonstrate to each tenant the RCCB and MCBs in operation and hand to each tenant a leaflet covering this operation		Item		6.00
		Carried forward		1,644.23

	Quantity	Unit	Rate	Total £
		Brought forward		1,644.23
BUILDER'S WORK IN CONNECTION WITH ELECTRICAL SERVICES				
Cut out in brick or block walls, and make good plaster for				
single socket outlet and/or fuse connection unit	5	each	5.35	26.75
twin socket outlet	5	each	6.50	32.50
low-level cooker connection unit and/or deep pattern socket box	2	each	5.35	10.70
Cut chase in brick or block wall, make good plaster 50mm wide				
cable chase	30	m	1.78	53.40
UPGRADING OF ELECTRICAL SERVICES TO ONE BEDROOM, TWO PERSON BUNGALOW				1,767.58

New electical installation

The work comprises the total electrical installation of high volume private properties. Rates, particularly cables, reflect work carried out on a large number of properties on one estate, therefore requiring large quantities of cables and accessories to be purchased which attract larger discounts. All work is installed flush, either 'fished' through dry linings or covered by suitably sized heavy duty plastic capping to protect the cables.

It has been assumed that the contractor is a member of the National Inspection Council for Electrical Instasllation Contracting and will carry out the wiring in accordance with the 16th edition of the Institute of Electrical Engineers' Wiring Regulations.

Each property is fitted with with a consumer unit with split load integral isolator suitably rated, RCCB, MCBs, bell transformers and blanks and all circuits are suitably labelled.

Each property is fitted with smoke detectors, complete with rechargeable batteries, each hard wired to the local lighting circuit and inter-connected. The detectors are ceiling mounted and located in the hall and first floor landings in the houses.

The installation has switches mounted 1200mm above finished floor level and socket outlets, except in the kitchen, 600mm above finished floor level. Outlets in kitchens are 200mm above worktop level generally.

Appliances housed beneath worktops, i.e. refrigerators, freezers, washing machines, water softeners and waste disposal units, have their socket outlets fitted 600mm above floor level and independently controlled by a switched fused connection unit, 200mm above worktop level.

Lighting is provided to all rooms as follows:

- pendant fittings to living rooms, dining rooms, landings, bedrooms and study areas
- single 1500mm fluorescent fitting with diffuser to kitchen forming composite part of kitchen and utility room
- batten fittings to halls, WCs and stores (under stairs) and garage
- zones 1-3 down lighters to bathrooms
- external bulkhead fitting adjacent to front door
- final connections to under cupboard kitchen unit lighting supplies as a composite part of unit
- two-way switches between halls and landings.

All earth bonding and cross-bonding within each property and main earth is installed. Each property is provided with the following as a minimum requirement. Other equipment or fittings, i.e. external IP rated socket outlet, wall light facility (wiring and switch point only) and additional socket outlets, switched, which may be added to the installation at the owner's request are to be treated as a variation to the contract.

Three bedroom, five person house; Type A

Ground floor

Living room

1 no. pendant light fitting
3 no. double socket outlets; switched
1 no. plate switche

Kitchen

1 no. fluorescent fitting
1 no. plate switch
4 no. double socket outlets; switched
1 no. cooker control unit with connection unit
1no. fuse connection unit for boiler
1 no. isolating switch for hob with connection unit
1 no. immersion switch with neon indicator
4 no. switched spurs
4 no. single socket outlets; switched
3 no. final connections to kitchen cupboard lights

Dining room

1 no. plate switch
1 no. single socket outlet; switched
3 no. double socket outlet; switched
1 no. pendant light fitting

Hall/external porch

2 no. batten light fittings
1 no. plate switch (hall, landing and outside light)
1 no. doublle socket outlet; switched
1 no. smoke alarm (interconnected)
1 no. bulkhead fitting; external

Store cupboard

1 no. batten light fitting
1 no. plate switch

First floor

Landing

1 no. pendant light fitting
1 no. plate switch
1 no. smoke alarm (interconnected)
1 no. single socket outlet; switched

Bedroom 1 (including en-suite)

3 no. double socket outlets, switched
1 no. pendant light fitting
1 no. plate switch
2 no. down light fittings (zones 1-3)

Bedroom 2

3 no. double socket outlets; switched
1 no. pendant light fitting
1 no. plate switch

Bedroom 3

2 no. double socket outlets; switched
1 no. pendant light fitting
1 no. plate switch

Bathroom

1 no. plate switch (external to bathroom)
3 no. down light fittings (zones 1-3)

Cylinder cupboard

1 no. double pole switched connection unit with flex outlet for immersion heater

Four bedroom, six person house; Type B

Ground floor

Living room

1 no. pendant light fitting
4 no. double socket outlets; switched
1 no. plate switch

Kitchen
1 no. fluorescent fitting
1 no. plate switch
4 no. double socket outlets; switched
1 no. fuse connection for boiler
1 no. cooker control unit with connection unit
1 no. isolating switch for hob with connection unit
1 no. immersion switch with neon indicator
4 no. switched spurs
4 no. single socket outlets; switched
4 no. final connections to kitchen cupboard lights

Dining room
1 no. plate switch
1 no. single socket outlet; switched
3 no. double socket outlets; switched
1 no. pendant light fitting

Hall/external porch
1 no. batten light fitting
1 no. plate switch (hall, landing and outside light)
1 no. single socket outlet; switched
1 no. smoke alarm (interconnected)
1 no. bulkhead fitting; external

Store cupboard
1 no. batten light fitting
1 no. plate switch

WC
1 no. batten lamp holder
1 no. plate switch (external to WC)

Study area
1 no. pendant light fitting
1 no. plate switch
3 no. double socket outlets; switched

Garage
2 no. plate switches
3 no. double socket outlets; switched
2 no. batten lamp holders

First floor

Landing
1 no. pendant light fitting
1 no. plate switch
1 no. smoke alarm (interconnected)
1 no. double socket outlet; switched

Bedroom 1 (including en-suite)
4 no. double socket outlets; switched
2 no. pendant light fittings
1 no. plate switch
2 no. down light fittings (zones 1-3)

Bedroom 2
3 no. double socket outlets; switched
1 no. pendant light fitting
1 no. plate switch

Bedroom 3
1 no. single socket outlet; switched
3 no. double socket outlets; switched
1 no. pendant light fitting
1 no. plate switch

Bathroom
1 no. plate switch (external to bathroom)
3 no. down light fittings (zones 1-3)

WC
1 no. batten lamp holder
1 no. plate switch (external to WC)

Cylinder cupboard
1 no. double pole switched connection unit with flex outlet for immersion heater

Five bedroom, seven person house; Type B

Ground floor
Living room
2 no. pendant light fittings
4 no. double socket outlets; switched
2 no. plate switches

Kitchen
1 no. fluorescent fitting
1 no. plate switch
5 no. double socket outlets; switched
1 no. cooker control unit with connection unit
1 no. isolating switch for hob with connection unit
1 no. immersion switch with neon indicator
3 no. switched spurs
3 no. single socket outlets; switched
4 no. final connections to kitchen cupboard lights

Utility room
1 no. fluorescent fitting
1 no. plate switch
3 no. double socket outlets; switched
3 no. switched spurs
3 no. single socket outlets; switched
1 no. switched spur to boiler

Dining room
2 no. plate switches
4 no. double socket outlet; switched
1 no. pendant light fitting

Porch/hall
1 no. pendant light fitting
1 no. batten light fitting
1 no. plate switch (porch/hall, landing and outside light)
2 no. double socket outlets; switched
1 no. smoke alarm (interconnected)
1 no. bulkhead fitting; external

Store cupboard
1 no. batten light fitting
1 no. plate switch

WC
1 no. batten lamp holder
1 no. plate switch (external to WC)

Study area
- 2 no. pendant light fittings
- 1 no. plate switch
- 5 no. double socket outlets; switched

Double garage
- 2 no. plate switches
- 5 no. double socket outlets; switched
- 4 no. batten lamp holders

First floor

Landing
- 1 no. pendant light fitting
- 1 no. plate switch
- 1 no. smoke alarm (interconnected)
- 1 no. double socket outlet; switched

Bedroom 1 (including en-suite)
- 1 no. single socket outlets; switched
- 3 no. double socket outlets; switched
- 2 no. pendant light fittings
- 2 no. plate switches
- 3 no. down light fittings (zones 1-3)

Bedroom 2
- 3 no. double socket outlets; switched
- 1 no. pendant light fitting
- 1 no. plate switch

Bedroom 3
- 3 no. double socket outlets; switched
- 2 no. pendant light fittings
- 1 no. plate switch

Bedroom 4
- 3 no. double socket outlets; switched
- 1 no. pendant light fitting
- 1 no. plate switch

Bedroom 5
- 3 no. double socket outlets; switched
- 1 no. pendant light fitting
- 1 no. plate switch

Bathroom 1
1 no. plate switch (external to bathroom)
3 no. down light fittings (zones 1-3)

Bathroom 2
1 no. plate switch (external to bathroom)
2 no. down light fittings (zones 1-3)

WC
1 no. batten lamp holder
1 no. plate switch (external to WC)

Cylinder cupboard
1 no. double pole switched connection unit with flex outlet for immersion heater

The external bulkhead fittings are manufactured by 'Thorn' and are fixed to the soffit of the porch. On completion of the work an NICEIC is produced and the Contractor demonstrates to the owner the operation of the RCCB and MCBs and leaves a written instructioncovering this operation for future use.

All the houses are two storey, built of traditional construction brick and block external walling with dry lining type internal "partitions. The ground floor is solid and the first floor is softwood boarding on timber joists.

	Quantity	Unit	Rate	Total £
NEW ELECTRICAL INSTALLATION OF THREE BEDROOM, FIVE PERSON HOUSE				
LV/HV CABLES AND WIRING				
600/1000 volt grade PVC insulated, PVC sheathed cable, single core				
Meter tails				
25mm2		Item		16.29
600/1000 volt grade PVC insulated, PVC sheathed cable, including PVC capping where necesary				
Drawn into voids or chases or clipped to backgrounds; twin and earth cabling				
1.5mm2	250	m	1.84	460.00
2.5mm2	175	m	2.14	374.50
6.0mm2	20	m	5.75	115.00
Drawn into voids or chases or clipped to backgrounds; three core and earth cabling				
1.5mm2	15	m	5.03	75.45
300/500 volt grade, PVC insulated, PVC sheathed heat resiatant circular cables; copper stranded conductors; BS 6141; in tails including termination at both ends				
Supply to immersion heater and boiler				
3 core; 2.5mm2	2	m	11.22	22.44
		Carried forward		1,063.68

	Quantity	Unit	Rate	Total £
		Brought forward		1,063.68
EARTHING AND BONDING COMPONENTS				
Copper earth connection including copper insulated cables and final connection and PVC capping where necessary				
Main earth between incoming main and consumer unit				
16mm2		Item		11.90
Supply and install main equipotential earth bonding between incoming mains and other incoming services including earth bonding clamps				
10mm2		Item		78.34
Supply and install cross-bonding to water services (sinks, boilers, etc.) including earth bonding clamps				
6mm2		Item		39.80
LV SWITCH GEAR AND DISTRIBUTION				
Consumer unit; split-load consumer unit complete with integral isolator, RCCB, MCBs; all suitably rated				
		Carried forward		1,193.72

	Quantity	Unit	Rate	Total £
		Brought forward		1,193.72
SP and N; 17 module split-load insulated consumer unit fitted with various MCBs, the remainder fitted with blanks; to backgrounds requiring fixings				
consumer unit	1	each	178.37	178.37
extra for				
6 amp, SP MCB	3	each	9.23	27.69
16 amp, SP MCB	1	each	8.85	8.85
32 amp, SP MCB	2	each	8.85	17.70
45 amp, SP MCB	1	each	9.23	9.23
bell transformer	1	each	20.52	20.52
LUMINAIRES AND LAMPS				
Surface-mounted internal luminaires				
White plastic ceiling rose, lamp holder complete with white flexible PVC insulated cable not exceeding 225mm in length; to backgrounds requiring fixings				
pendant	6	each	7.70	46.20
White plastic batten lamp holder complete with skirt; to backgrounds requiring fixings				
ceiling lamp holder	3	each	7.81	23.43
Fluorescent luminaire complete with lamp and perspex diffuser; to backgrounds requiring fixings				
1500mm long; single tube	1	each	26.77	26.77
		Carried forward		1,552.48

	Quantity	Unit	Rate	Total £
		Brought forward		1,552.48
Recessed mains voltage luminaires IP44 rated complete with opal lens and lamp				
60 watt tungsten downlighter, zones 1-3	5	each	18.10	90.50
Final conections to to under cupboard lighting supplied by others				
229mm long; tungsten	3	each	1.54	4.62
ACCESSORIES FOR ELECTRICAL SERVICES				
250 volt grade flush-mounted accessories including dry lining back box; fitted into aperture formed by others				
White plastic switch plates				
1 gang; 1 way; single pole	8	each	4.63	37.04
2 gang; 2 way; single pole	1	each	6.75	6.75
3 gang; 2 way; single pole	1	each	13.86	13.86
1 gang; 2 way; single pole	1	each	5.37	5.37
Switch socket outlets				
13 amp; 1 gang	5	each	7.39	36.95
13 amp; 2 gang	19	each	10.27	195.13
Switch fuse indicator connection unit including outgoing final connections				
13amp; double pole;	5	each	13.19	65.95
		Carried forward		2,008.65

	Quantity	Unit	Rate	Total £
		Brought forward		2,008.65
Switch fuse connection unit including outgoing final connections engraved 'Immersion heater'				
13 amp; double pole; including neon indicator	1	each	18.67	18.67
Low-level cooker connection unit including final connections				
connection unit	2	each	13.40	26.80
Double pole indicated switch				
20 amp; double pole	2	each	17.00	34.00
White plastic cooker control unit				
50 amp; cooker control unit	1	each	19.77	19.77
Sundry work				
Door chimes and bell push including associated wiring and protection; to backgrounds requiring fixings				
door bell system	1	each	36.98	36.98
250/9 volt grade surface-mounted smoke detector and fire alarm unit complete with rechargeable battery for back-up purposes; hard wired; to backgrounds requiring fixings				
		Carried forward		2,144.87

	Quantity	Unit	Rate	Total £
		Brought forward		2,144.87
Smoke detector/fire alarm unit complete with associated PVC insulated, PVC sheathed cable taken from local lighting circuit				
smoke detector/fire alarm unit	2	each	33.59	67.18
EXTERNAL LIGHTING				
FITTINGS AND ACCESSORIES				
250 volt grade tungsten luminaire complete with lamp; as manufactured by 'Thorn'				
Surface-mounted bulkhead luminaire including conduit sleeve through wall, bushing; final connections and the like; to backgrounds requiring fixings				
'Thorn' bulkhead luminaire	1	each	42.66	42.66
Testing				
Final testing as per IEE Wiring Regulations and production of standard NICEIC documentation per dwelling				
three bedroom, five person house		Item		42.00
Operating and training				
Demonstrate to each owner the RCCB and MCBs in operation and hand to each owner a leaflet covering this operation leaflet covering this operation		Item		6.00
UPGRADING OF ELECTRICAL SERVICES TO THREE BEDROOM, FIVE PERSON HOUSE				2,302.71

	Quantity	Unit	Rate	Total £
NEW ELECTRICAL INSTALLATION OF FOUR BEDROOM, SIX PERSON HOUSE				
LV/HV CABLES AND WIRING				
600/1000 volt grade PVC insulated, PVC sheathed cable, single core				
Meter tails				
25mm2		Item		16.29
600/1000 volt grade PVC insulated, PVC sheathed cable, including PVC capping where necesary				
Drawn into voids or chases or clipped to backgrounds; twin and earth cabling				
1.5mm2	300	m	1.84	552.00
2.5mm2	250	m	2.14	535.00
6.0mm2	30	m	5.75	172.50
Drawn into voids or chases or clipped to backgrounds; three core and earth cabling				
1.5mm2	25	m	5.03	125.75
300/500 volt grade, PVC insulated, PVC sheathed heat resiatant circular cables; copper stranded conductors; BS 6141; in tails including termination at both ends				
Supply to immersion heater and boiler				
3 core; 2.5mm2	2	m	11.22	22.44
		Carried forward		1,423.98

	Quantity	Unit	Rate	Total £
		Brought forward		1,423.98
EARTHING AND BONDING COMPONENTS				
Copper earth connection including copper insulated cables and final connection and PVC capping where necessary				
Main earth between incoming main and consumer unit				
16mm2		Item		11.90
Supply and install main equipotential earth bonding between incoming mains and other incoming services including earth bonding clamps				
10mm2		Item		78.34
Supply and install cross-bonding to water services (sinks, boilers, etc.) including earth bonding clamps				
6mm2		Item		39.80
LV SWITCH GEAR AND DISTRIBUTION				
Consumer unit; split-load consumer unit complete with integral isolator, RCCB, MCBs; all suitably rated				
		Carried forward		1,554.02

	Quantity	Unit	Rate	Total £
		Brought forward		1,554.02
SP and N; 17 module split-load insulated consumer unit fitted with various MCBs, the remainder fitted with blanks; to backgrounds requiring fixings				
consumer unit	1	each	178.37	178.37
extra for				
6 amp, SP MCB	3	each	9.23	27.69
16 amp, SP MCB	1	each	8.85	8.85
32 amp, SP MCB	2	each	8.85	17.70
45 amp, SP MCB	1	each	9.23	9.23
bell transformer	1	each	20.52	20.52
LUMINAIRES AND LAMPS				
Surface-mounted internal luminaires				
White plastic ceiling rose, lamp holder complete with white flexible PVC insulated cable not exceeding 225mm in length; to backgrounds requiring fixings				
pendant	8	each	7.70	61.60
White plastic batten lamp holder complete with skirt; to backgrounds requiring fixings				
ceiling lamp holder	6	each	7.81	46.86
Fluorescent luminaire complete with lamp and perspex diffuser; to backgrounds requiring fixings				
1500mm long; single tube	1	each	26.77	26.77
		Carried forward		1,951.61

	Quantity	Unit	Rate	Total £
		Brought forward		1,951.61
Recessed mains voltage luminaires IP44 rated complete with opal lens and lamp				
60 watt tungsten downlighter, zones 1-3	5	each	18.10	90.50
Final conections to to under cupboard lighting supplied by others				
229mm long; tungsten	4	each	1.54	6.16
ACCESSORIES FOR ELECTRICAL SERVICES				
250 volt grade flush-mounted accessories including dry lining back box; fitted into aperture formed by others				
White plastic switch plates				
1 gang; 1 way; single pole	9	each	4.63	41.67
2 gang; 2 way; single pole	2	each	6.75	13.50
3 gang; 2 way; single pole	1	each	13.86	13.86
1 gang; 2 way; single pole	1	each	5.37	5.37
Switch socket outlets				
13 amp; 1 gang	7	each	7.39	51.73
13 amp; 2 gang	25	each	10.27	256.75
Switch fuse indicator connection unit including outgoing final connections				
13amp; double pole;	5	each	13.19	65.95
		Carried forward		2,497.10

	Quantity	Unit	Rate	Total £
		Brought forward		2,497.10
Switch fuse connection unit including outgoing final connections engraved 'Immersion heater'				
13 amp; double pole; including neon indicator	1	each	18.67	18.67
Low-level cooker connection unit including final connections				
connection unit	2	each	13.40	26.80
Double pole indicated switch				
20 amp; double pole	2	each	17.00	34.00
White plastic cooker control unit				
50 amp; cooker control unit	1	each	19.77	19.77
250 volt grade surface-mounted accessories including moulded plastic back boxes; to backgrounds requiring fixings (garage)				
White plastic switch plate				
1 gang; 2way; single pole	2	each	7.67	15.34
White plastic switch socket outlets				
13 amp; 2 gang	3	each	12.41	37.23
		Carried forward		2,648.91

	Quantity	Unit	Rate	Total £
		Brought forward		2,648.91
Sundry work				
Door chimes and bell push including associated wiring and protection; to backgrounds requiring fixings				
door bell system	1	each	36.98	36.98
250/9 volt grade surface-mounted smoke detector and fire alarm unit complete with rechargeable battery for back-up purposes; hard wired; to backgrounds requiring fixings				
Smoke detector/fire alarm unit complete with associated PVC insulated, PVC sheathed cable taken from local lighting circuit				
smoke detector/fire alarm unit	2	each	33.59	67.18
EXTERNAL LIGHTING				
FITTINGS AND ACCESSORIES				
250 volt grade tungsten luminaire complete with lamp; as manufactured by 'Thorn'				
Surface-mounted bulkhead luminaire including conduit sleeve through wall, bushing; final connections and the like; to backgrounds requiring fixings				
'Thorn' bulkhead luminaire	1	each	42.66	42.66
		Carried forward		2,795.73

	Quantity	Unit	Rate	Total £
		Brought forward		2,795.73
250 volt grade surface-mounted weatherproof accessorties including polycarbonate back box; to backgrounds requiring fixings				
13 amp; 1 gang	2	each	31.14	62.28
Testing				
Final testing as per IEE Wiring Regulations and production of standard NICEIC documentation per dwelling				
four bedroom, six person house		Item		48.00
Operating and training				
Demonstrate to each owner the RCCB and MCBs in operation and hand to each owner a leaflet covering this operation leaflet covering this operation		Item		6.00
UPGRADING OF ELECTRICAL SERVICES TO FOUR BEDROOM, SIX PERSON HOUSE				2,912.01

	Quantity	Unit	Rate	Total £
NEW ELECTRICAL INSTALLATION OF FIVE BEDROOM, SEVEN PERSON HOUSE				
LV/HV CABLES AND WIRING				
600/1000 volt grade PVC insulated, PVC sheathed cable, single core				
Meter tails				
25mm2		Item		16.29
600/1000 volt grade PVC insulated, PVC sheathed cable, including PVC capping where necesary				
Drawn into voids or chases or clipped to backgrounds; twin and earth cabling				
1.5mm2	400	m	1.84	736.00
2.5mm2	300	m	2.14	642.00
6.0mm2	30	m	5.75	172.50
Drawn into voids or chases or clipped to backgrounds; three core and earth cabling				
1.5mm2	35	m	5.03	176.05
300/500 volt grade, PVC insulated, PVC sheathed heat resiatant circular cables; copper stranded conductors; BS 6141; in tails including termination at both ends				
Supply to immersion heater and boiler				
3 core; 2.5mm2	2	m	11.22	22.44
		Carried forward		1,765.28

	Quantity	Unit	Rate	Total £
		Brought forward		1,765.28
EARTHING AND BONDING COMPONENTS				
Copper earth connection including copper insulated cables and final connection and PVC capping where necessary				
Main earth between incoming main and consumer unit				
16mm2		Item		11.90
Supply and install main equipotential earth bonding between incoming mains and other incoming services including earth bonding clamps				
10mm2		Item		78.34
Supply and install cross-bonding to water services (sinks, boilers, etc.) including earth bonding clamps				
6mm2		Item		39.80
LV SWITCH GEAR AND DISTRIBUTION				
Consumer unit; split-load consumer unit complete with integral isolator, RCCB, MCBs; all suitably rated				
		Carried forward		1,895.32

	Quantity	Unit	Rate	Total £
		Brought forward		1,895.32
SP and N; 17 module split-load insulated consumer unit fitted with various MCBs, the remainder fitted with blanks; to backgrounds requiring fixings				
consumer unit	1	each	178.37	178.37
extra for				
6 amp, SP MCB	3	each	9.23	27.69
16 amp, SP MCB	1	each	8.85	8.85
32 amp, SP MCB	2	each	8.85	17.70
45 amp, SP MCB	1	each	9.23	9.23
bell transformer	1	each	20.52	20.52
LUMINAIRES AND LAMPS				
Surface-mounted internal luminaires				
White plastic ceiling rose, lamp holder complete with white flexible PVC insulated cable not exceeding 225mm in length; to backgrounds requiring fixings				
pendant	14	each	7.70	107.80
White plastic batten lamp holder complete with skirt; to backgrounds requiring fixings				
ceiling lamp holder	2	each	7.81	15.62
Fluorescent luminaire complete with lamp and perspex diffuser; to backgrounds requiring fixings				
1500mm long; single tube	2	each	26.77	53.54
		Carried forward		2,334.64

	Quantity	Unit	Rate	Total £
		Brought forward		2,334.64
Recessed mains voltage luminaires IP44 rated complete with opal lens and lamp				
60 watt tungsten downlighter, zones 1-3	8	each	18.10	144.80
Final conections to to under cupboard lighting supplied by others				
229mm long; tungsten	4	each	1.54	6.16
ACCESSORIES FOR ELECTRICAL SERVICES				
250 volt grade flush-mounted accessories including dry lining back box; fitted into aperture formed by others				
White plastic switch plates				
1 gang; 1 way; single pole	10	each	4.63	46.30
2 gang; 2 way; single pole	1	each	6.75	6.75
3 gang; 2 way; single pole	1	each	13.86	13.86
1 gang; 2 way; single pole	5	each	5.37	26.85
Switch socket outlets				
13 amp; 1 gang	7	each	7.39	51.73
13 amp; 2 gang	39	each	10.27	400.53
Switch fuse indicator connection unit including outgoing final connections				
13amp; double pole;	7	each	13.19	92.33
		Carried forward		3,123.95

	Quantity	Unit	Rate	Total £
		Brought forward		3,123.95
Switch fuse connection unit including outgoing final connections engraved 'Immersion heater'				
13 amp; double pole; including neon indicator	1	each	18.67	18.67
Low-level cooker connection unit including final connections				
connection unit	2	each	13.40	26.80
Double pole indicated switch				
20 amp; double pole	2	each	17.00	34.00
White plastic cooker control unit				
50 amp; cooker control unit	1	each	19.77	19.77
250 volt grade surface-mounted accessories including moulded plastic back boxes; to backgrounds requiring fixings (garage)				
White plastic switch plate				
1 gang; 2way; single pole	2	each	7.67	15.34
White plastic switch socket outlets				
13 amp; 2 gang	5	each	12.41	62.05
		Carried forward		3,300.58

	Quantity	Unit	Rate	Total £
		Brought forward		3,300.58
TRUNKING				
UPVC extra super high impact grade 'Mini' trunking; to backgrounds requiring fixings				
Single compartment				
16 × 25mm	14	m	4.92	68.88
50 × 25mm	10	m	7.37	73.70
Extra over				
50 × 25mm; equal tees	8	each	3.31	26.48
50 × 25mm; 90° bends	2	each	3.14	6.28
50 × 25mm; end caps	2	each	1.54	3.08
16 × 25mm; couplers	7	each	1.06	7.42
Sundry work				
Door chimes and bell push including associated wiring and protection; to backgrounds requiring fixings				
door bell system	1	each	36.98	36.98
250/9 volt grade surface-mounted smoke detector and fire alarm unit complete with rechargeable battery for back-up purposes; hard wired; to backgrounds requiring fixings				
Smoke detector/fire alarm unit complete with associated PVC insulated, PVC sheathed cable taken from local lighting circuit				
smoke detector/fire alarm unit	2	each	33.59	67.18
		Carried forward		3,590.58

	Quantity	Unit	Rate	Total £
		Brought forward		3,590.58
EXTERNAL LIGHTING				
FITTINGS AND ACCESSORIES				
250 volt grade tungsten luminaire complete with lamp; as manufactured by 'Thorn'				
Surface-mounted bulkhead luminaire including conduit sleeve through wall, bushing; final connections and the like; to backgrounds requiring fixings				
'Thorn' bulkhead luminaire	1	each	42.66	42.66
250 volt grade surface-mounted weatherproof accessorties including polycarbonate back box; to backgrounds requiring fixings				
13 amp; 1 gang	2	each	31.14	62.28
Testing				
Final testing as per IEE Wiring Regulations and production of standard NICEIC documentation per dwelling				
five bedroom, seven person house		Item		42.00
Operating and training				
Demonstrate to each owner the RCCB and MCBs in operation and hand to each owner a leaflet covering this operation leaflet covering this operation		Item		6.00
UPGRADING OF ELECTRICAL SERVICES TO FIVE BEDROOM, SEVEN PERSON HOUSE				3,700.86

Part Three

BUSINESS MATTERS

Starting a business

Running a business

Taxation

Starting a business

Most small businesses come into being for one of two reasons – ambition or desperation! A person with genuine ambition for commercial success will never be completely satisfied until he has become self-employed and started his own business. But many successful businesses have been started because the proprietor was forced into this course of action because of redundancy.

Before giving up his job, the would-be businessman should consider carefully whether he has the required skills and the temperament to survive in the highly competitive self-employed market. Before commencing in business it is essential to assess the commercial viability of the intended business because it is pointless to finance a business that is not going to be commercially viable.

In the early stages it is important to make decisions such as: What exactly is the product being sold? What is the market view of that product? What steps are required before the developed product is first sold and where are those sales coming from?

As much information as possible should be obtained on how to run a business before taking the plunge. Sales targets should be set and it should be clearly established how those important first sales are obtained. Above all, do not underestimate the amount of time required to establish and finance a new business venture.

Whatever the size of the business it is important that you put in writing exactly what you are trying to do. This means preparing a business plan that will not only assist in establishing your business aims but is essential if you need to raise finance. The contents of a typical business plan are set out later. It is important to realise that you are not on your own and there are many contacts and advising agencies that can be of assistance.

Potential customers and trade contacts

Many persons intending to start a business in the construction industry will have already had experience as employees. Use all contacts to check the market, establish the sort of work that is available and the current charge-out rates.

In the domestic market, check on the competition for prices and services provided. Study advertisements for your kind of work and try to get firm promises of work before the start-up date.

Testing the market

Talk to as many traders as possible operating in the same field. Identify if the market is in the industrial, commercial, local government or in the domestic field. Talk to prospective customers and clients and consider how you can improve on what is being offered in terms of price, quality, speed, convenience, reliability and back-up service.

Business links

There is no shortage of information about the many aspects of starting and running your own business. Finance, marketing, legal requirements, developing your business idea and taxation matters are all the subject of a mountain of books, pamphlets, guides and courses so it should not be necessary to pay out a lot of money for this information. Indeed, the likelihood is that the aspiring businessman will be overwhelmed with information and will need professional guidance to reduce the risk of wasting time on studying unnecessary subjects.

Business Links are now well established and provide a good place to start for both information and advice. These organisations provide a 'one-stop-shop' for advice and assistance to owner-managed businesses. They will often replace the need to contact Training and Enterprise Councils (TECs) and many of the other official organisations listed below.
Point of contact: telephone directory for address.

Training and Enterprise Councils (TECs)

TECs are comprised of a board of directors drawn from the top men in local industry, commerce, education, trade unions etc., who, together with their staff and experienced business counsellors, assist both new and established concerns in all aspects of running a business. This takes the form of across-the-table advice and also hands-on assistance in management, marketing and finance if required. There are also training courses and seminars available in most areas together with the possibility of grants in some areas.
Point of contact: local Jobcentre or Citizens' Advice Bureau.

Banks

Approach banks for information about the business accounts and financial services that are available. Your local Business Link can advise on how best to find a suitable bank manager and inform you as to what the bank will require.

Shop around several banks and branches if you are not satisfied at first because managers vary widely in their views on what is a viable business proposition. Remember, most banks have useful free information packs to help business start-up.

Point of contact: local bank manager.

HM Inspector of Taxes

Make a preliminary visit to the local tax office enquiry counter for their publications on income tax and national insurance contributions.

SA/Bk 3	Self assessment. A guide to keeping records for the self employed
IR 15(CIS)	Construction Industry Tax Deduction Scheme
CWL	Starting your own business,
IR 40(CIS)	Conditions for Getting a Sub-Contractor's Tax Certificate
NE1	PAYE for Employers (if you employ someone)
NE3	PAYE for new and small Employers
IR 56/N139	Employed or Self-Employed. A guide for tax and National Insurance
CA02	National Insurance contributions for self employed people with small earnings.

Remember, the onus is on the taxpayer, within three months, to notify the Inland Revenue that he is in business and failure to do so may result in the imposition of £100 penalty. Either send a letter or use the form provided at the back of the *'Starting your own business booklet'* to the Inland Revenue National Insurance Contributions Office and they will inform your local tax office of the change in your employment status.

Point of contact: telephone directory for address.

Inland Revenue National Insurance Contributions Office

Self Employment Services
Customer Accounts Section
Longbenton
Newcastle NE 98 1ZZ

Telephone the Call Centre on 0845 9154655 and ask for the following publications:

CWL2	Class 2 and Class 4 Contributions for the Self Employed
CA02	People with Small Earnings from Self-Employment
CA04	Direct Debit - The Easy Way to Pay. Class 2 and Class 3
CA07	Unpaid and Late Paid Contributions

and for Employers

CWG1	Employer's Quick Guide to PAYE and NIC Contributions
CA30	Employer's Manual to Statutory Sick Pay

VAT

The VAT office also offer a number of useful publications, including;

700	The VAT Guide
700/1	Should I be Registered for VAT?
731	Cash Accounting
732	Annual Accounting
742	Land and Property

Information about the Cash Accounting Scheme and the introduction of annual VAT returns are dealt with later.
Point of contact: telephone directory for address.

Local authorities

Authorities vary in provisions made for small businesses but all have been asked to simplify and cut delays in planning applications. In Assisted Areas, rent-free periods and reductions in rates may be available on certain industrial and commercial properties. As a preliminary to either purchasing or renting business premises, the following booklets will be helpful:

Step by Step Guide to Planning Permission for Small Businesses, and *Business Leases and Security of Tenure*

Both are issued by the Department of Employment and are available at council offices, Citizens' Advice Bureaux and TEC offices. Some authorities run training schemes in conjunction with local industry and educational establishments.

Point of contact: usually the Planning Department - ask for the Industrial Development or Economic Development Officer.

Department of Trade and Industry

The services formally provided by the Department are now increasingly being provided by Business Link . The Department can still, however, provide useful information on available grants for start-ups.
Point of contact: telephone 0207-215 5000 and ask for the address and telephone number of the nearest DTI office and copies of their explanatory booklets.

Department of Transport and the Regions

Regulations are now in force relating to all forms of waste other than normal household rubbish. Any business that produces, stores, treats, processes, transports, recycles or disposes of such waste has a 'duty of care' to ensure it is properly discarded and dealt with.

Practical guidance on how to comply with the law (it is a criminal offence punishable by a fine not to) is contained in a booklet *Waste Management: The Duty of Care: A Code of Practice,* obtainable from HMSO Publication Centre, PO Box 276, London SW8 5DT. Telephone 0207-873 9090.

Accountant

The services of an accountant are to be strongly recommended from the beginning because the legal and taxation requirements start immediately and must be properly complied with if trouble is to be avoided later. A qualified accountant must be used if a limited company is being formed but an accountant will give advice on a whole range of business issues including book-keeping, tax planning and compliance to finance raising and will help in preparing annual accounts.

It is worth spending some time finding an accountant who has other clients in the same line of business and is able to give sound advice particularly on taxation and business finance and is not so overworked that damaging delays in producing accounts are likely to arise. Ask other traders whether they can recommend their own accountant. Visit more than one firm of accountants, ask about the fees they charge and how much the production of annual accounts and agreement with the Inland Revenue are likely to cost. A good accountant is worth every penny of his fees and will save you both money and worry.

Solicitor

Many businesses operate without the services of a solicitor but there are a number of occasions when legal advice should be sought. In particular, no-one should sign a lease of premises without taking legal advice because a business can encounter financial difficulty through unnoticed liabilities in its lease. Either an accountant or solicitor will help with drawing up a partnership agreement which all partnerships should have. A solicitor will also help to explain complex contractual terms and prepare draft contracts if the type of business being entered into requires them.

Insurance broker

Policies are available to cover many aspects of business including:

- employer's liability - compulsory if the business has employees
- public liability - essential in the construction industry
- motor vehicles
- theft of stock, plant and money
- fire and storm damage
- personal accident and loss of profits
- keyman cover.

Brokers are independent advisers who will obtain competitive quotations on your behalf. See more than one broker before making a decision - their advice is normally given free and without obligation.

Point of contact: telephone directory or write for a list of local members to:

The British Insurance Brokers' Association
Consumer Relations Department
BIBA House
14 Bevis Marks
London
EC3A 7NT (telephone: 0207-623 9043)
or contact
The Association of British Insurers
51 Gresham Street
London
EC2V 7HQ (telephone: 0207-600 3333)

who will supply free a package of very useful advice files specially designed for the small business.

The Health and Safety Executive

The Executive operates the legislation covering everyone engaged in work activities and has issued a very useful set of '*Construction Health Hazard Information Sheets*' covering such topics as handling cement, lead and solvents, safety in the use of ladders, scaffolding, hoists, cranes, flammable liquids, asbestos, roofs and compressed gases etc. A pack of these may be obtained free from your local HSE office or The Health & Safety Executive Central Office, Sheffield (telephone: (01142-892345) or HSE Publications (telephone: 01787-881165).

Business plan

As stated before, once the relevant information has been obtained it should be consolidated into a formal business plan. The complexity of the plan will depend in the main on the size and nature of the business concerned. Consideration should be given to the following points.

Objectives

It is important to establish what you are trying to achieve both for you and the business. A provider of finance may be particularly influenced by your ability to achieve short- and medium-term goals and may have confidence in continuing to provide finance for the business. From an individual point of view, it is important to establish goals because there is little point in having a business that only serves to achieve the expectations of others whilst not rewarding the would-be businessman.

History

If you already own an existing business then commentary on its existing background structure and history to date can be of assistance. There is no substitute for experience and any existing contacts you have in the construction industry will be of assistance to you. The following points should also be considered for inclusion:

- a brief history of the business identifying useful contacts made
- the development of the business, highlighting significant successes and
  their relevance to the future
- principal reasons for taking the decision to pursue this new venture
- details of present financing of the business.

Products or services

It is important to establish precisely what it is you are going to sell. Does the product or service have any unique qualities which gives it your advantages over competitors? For example, do you have an ability to react more quickly than your competitors and are you perceived to deliver a higher quality product or service?
A typical business plan would include:

- description of the main products and services
- statement of disadvantages and advising how they will be overcome
- details of new products and estimated dates of introduction
- profitability of each product
- details of research and development projects
- after-sales support.

Markets and marketing strategy

This section of the business plan should show that thought has been given to the potential of the product. In this regard it can often be useful to identify major
competitors and make an overall assessment of their strengths and weaknesses, including the following:

- an overall assessment of the market, setting out its size and growth potential
- a statement showing your position within the market
- an identification of main customers and how they compare
- details of typical orders and buying habits
- pricing strategy
- anticipated effect on demand of pricing
- expectation of price movement
- details of promotions and advertising campaigns.

It is important to identify your customers and why they might buy from you. Those entering the domestic side of the business will need to think about the best way to reach potential customers. Are local word-of-mouth recommendations enough to provide reasonable work continuity. If not, what is the most effective method of advertising to reach your customer base?

Remember, advertising is costly. It is a waste of funds to place an advertisement in a paper circulating in areas A, B, C & D if the business only covers area A.

Research and development

If you are developing a product or a particular service, then an assessment should be made on what stage it is at and what further finance is required to complete it. It may also be useful to make an assessment on the vulnerability of the product or service to innovations being initiated by others.

Basis of operation

Detail what facilities you will require in order to carry on your trade in the form of property, working and storage areas, office space, etc. An assessment should also he made on the assistance you will require from others. Your business plan might include:

- a layman's guide to the process or work
- details of facilities, buildings and plant
- key factors affecting production, such as yields and wastage
- raw material demand and usage.

Management

This section is one of the most important because it demonstrates the capability of the would-be businessman. The skills you need will cover production, marketing, finance and administration. In the early stages you may be able to do this yourself but as the business grows it may be required to develop a team to handle these matters. The following points should be considered for inclusion in the plan:

- set out age, experience and achievements
- state additional management requirements in the future and how they are to be met
- identify current weaknesses and how they will be overcome
- state remuneration packages and profit expectations
- give detailed CVs in appendices.

Advertising and retraining may be required in order to identify and provide suitable personnel where expertise and experience are lacking.

Financial information

It is important to detail, if any, the present financial position of your business and the budgeted profit and loss accounts, cash flows and balance sheets. These integrated forecasts should be prepared for the next twelve months at monthly intervals and annually for the following two years.

If the forecasts are to be reasonably accurate then the businessman must make some early decisions about:

- the premises where the business will be based, the initial repairs and alterations that might he required and an assessment of the total cost

- which plant, equipment and transport are needed, whether they are to be leased or purchased and what the cost will be?
- how much stock of materials, if any, should be carried? - the bare minimum only should be acquired, so reliable suppliers should be found
- what will be the weekly bills for overheads, wages and the proprietor's living costs?
- what type of work is going to be undertaken, and how much profit can realistically be obtained?
- how often are invoices to be presented?

Your business plan should include the following information:

- explanation of how sales forecasts are prepared
- levels of production
- details of major variable overheads and estimates
- assumptions in cash flow forecasting, inflation and taxation.

Finance required and its application

The financial details given above should produce an accurate assessment of the funds required to finance the business. It is important to distinguish between those items that require permanent finance and those that will eventually be converted to cash because it is not usually advisable to finance long-term assets with personal equity.

Working capital such as stock and debtors can usually be obtained by an overdraft arrangement but your accountant or bank will advise you on this.

Executive summary

Although it is prepared last, this summary will be the first part of your business plan. Remember that business plans are prepared for busy people and their decision on finance may be based solely on this section. It should cover two or three pages and deal with the most important aspects and opportunities in your plan. Here are some of the main headings:

- key strategies
- finance required and how it is to be used
- management experience

- anticipated returns and profits
- markets.

The appendices should include:

- CVs of key personnel
- organisation charts
- market studies
- product advertising literature
- professional references
- financial forecasts
- glossary of terms.

If you feel that any additional information should be provided in support of your proposal, then this is usually best included in the appendices.

Follow up

Please remember that once your plan is prepared, it is important to re-examine it regularly and update the forecasts and financial information. This is a working document and can be an important tool in running the business.

Sources of finance

Personal funds

Finance, like charity, often begins at home and a would-be businessman should make a realistic assessment of his net worth, including the value of his house after deducting the mortgage(s) outstanding on it, savings, any car or van owned and any sums which the family are prepared to contribute but deducting any private borrowings which will come due for payment. The whole of these funds may not be available (for instance, money which has been loaned to a friend or relative who is known to be unable to repay at the present time).

It may not be desirable that all capital should be put at risk on a business venture so the following should be established:

- how much cash you propose to invest in the business
- whether the family home will be made available for any business borrowing
- state total finance required
- how finance is anticipated being raised
- interest and security to be provided
- expected return on investment.

Whilst it may be wise not to pledge too much of the family assets, it has to be remembered that the bank will be looking closely at the degree to which the proprietor has committed himself to the venture and will not be impressed by an application for a loan where the applicant is prepared to risk only a small fraction of his own resources.

Having decided how much of his own funds to contribute, the businessman can now see the level of shortfall and consider how best to fill it. Consideration should be given to partners where the shortfall is large and particularly when there is a need for heavy investment in fixed assets, such as premises and capital equipment. It may be worthwhile starting a limited company with others also subscribing capital and to allow the banks to take security against the book debts.

Banks

The first outside source of money to which most businessmen turn is the bank and here are a few guidelines on approaching a bank manager:

- present your business plan to him; remember to use conservative estimates which tend to understate rather than overstate the forecast sales and profits
- know the figures in detail and do not leave it to your accountant to explain them for you. The bank manager is interested in the businessman not his advisers and will be impressed if the businessman demonstrates a grasp of the financing of his business
- understand the difference between short- and long-term borrowing

- ask about the Government Loan Guarantee Scheme if there is a shortage of security for loans. The bank may be able to assist, or depending on certain conditions being met, the Government may guarantee a certain percentage of the bank loan.

Remember the bank will want their money back, so bank borrowings are usually required to be secured by charges on business assets. In start-up situations, personal guarantees from the proprietors are normally required. Ensure that if these are given they are regularly reviewed to see if they are still required.

Enterprise Investment Scheme - business angels

If an outside investor is sought in a business he will probably wish to invest within the terms of the Enterprise Investment Scheme which enables him to gain income tax relief at 20% on the amount of his investment. Additionally, any investment can be used to defer capital gains tax. The rules are complex and professional advice should always be sought.

Hire purchase/leasing

It is not always necessary to purchase assets outright that are required for the business and leasing and hire purchase can often form an integral part of a business's medium-term finance strategy.

Venture capital

In addition, there are a number of other financial institutions in the venture capital market that can help well-established businesses, usually limited companies, who wish to expand. They may also assist well-conceived start-ups. They will provide a flexible package of equity and loan capital but only for large amounts, usually sums in excess of £150,000 and often £250,000.

Usually the deal involves the financial institution having a minority interest in the voting share capital and a seat on the board of the company. Arrangements for the eventual purchase of the shares held by the finance company by the private shareholders are also normally incorporated in the scheme.

The Royal Jubilee and Princes Trust

These trusts through the Youth Business Initiative provide bursaries of not more than £1,000 per individual to selected applicants who are unemployed and age 25 or over. Grants may be used for tools and equipment, transport, fees, insurance, instruction and training but not for working capital, rent and rates, new materials or stock. They operate through a local representative whose name and address may be ascertained by contacting the Prince's Youth Business.
Point of contact: telephone 0207-321 6500.

The Business Start-up Scheme

This is an allowance of £50 per week, in addition to any income made from your business, paid for twenty weeks. To qualify you must be at least 18 and under 65, work at least 36 hours per week in the business and have been unemployed for at least six months or fall into one of the other categories: disabled, ex-HMS or redundant.

The first step is to get the booklet on the subject from your local Jobcentre or TEC that includes details on how and where to apply. Once in receipt of the enterprise allowance, you will also have the benefit of advice and assistance from an experienced businessman from your TEC. All the initial counselling services and training courses are free.

Running a business

Many businesses are run without adequate information being available to check trend in their vital areas, e.g. marketing, money and managerial efficiency. It is essential to look critically at all aspects of the business in order to maximise profits and reduce inefficiency. Regular meaningful information is required on which management can concentrate. This will vary according to the proprietor's business but will often concentrate on debtors, creditors, cash, sales and orders.

Proprietors often have the feeling that the business should be 'doing better' but are unable to identify what is going wrong. Sometimes there is the worrying phenomenon of a steadily increasing work programme coupled with a persistently reducing bank balance or rising overdraft. Some useful ways of checking the position and of identifying problem areas are given below.

Marketing

Throughout his business life the entrepreneur should continuously study the methods and approach of his competitors. A shortcoming frequently found in ailing concerns is that the proprietor thinks he knows what his customers want better than they do.

The term 'market research' sounds both difficult and expensive but a very simple form of it can be done quite effectively by the businessman and his sales staff. Existing and prospective customers should be approached and asked what they want in terms of price, quality, design, payment terms, follow-up service, guarantees and services.

The initial approach might be by a leaflet or letter followed by a personal call. As an on-going part of management, all staff with customer contact should be encouraged to enquire about and record customer preferences, complaints, etc. and feed it back to management.

Other sources of information can be trade and business journals, trade exhibitions, suppliers and representatives from which information about trends, new techniques and products can be obtained and studied. Valuable information can also be gained from studying competitors and the following questions should be asked:

- what do they sell and at what prices?

- what inducements do they offer to their customers, e.g. credit facilities, guarantees, free offers and discounts?
- how do they reach their customers - local/national advertising, mail shots, salesmen, local radio and TV?
- what are the strongest aspects of their appeal to customers and have they any weaknesses?

The businessman should apply all the information gathered from customers and competitors to his own services with a view to making sure he is offering the right product at the right price in the most attractive way and in the most receptive market.

In a small business where the proprietor is also his own salesman he must give careful thought on how he can best present his product and himself. For instance, if he is working solely within the construction industry his main problems are likely to centre on getting a C1S6 Certificate and using trade contacts to get sub-contract work.

However, for those who serve the general public, presentation can be a vital element in getting work. The customer is looking for efficiency, reliability and honesty in a trader and quality, price and style in the product. To bring out these facets in discussion with a potential customer is a skilled task. A short course on marketing techniques could pay handsome dividends. The Business Link will give the names and addresses of such courses locally.

Financial control

Unfortunately, some unsuccessful firms do not seek financial advice until too late when the downward trend cannot be halted. Earlier attention to the problems may have saved some of them so it is important to recognise the tell-tale signs. There are some tests and checks that can be done quite easily.

Cash flow

Cash flow is the lifeblood of the business and more businesses fail through lack of cash than for any other reason. Cash is generated through the conversion of work into debtors and then into payment and also through the deferral of the payment of supplies for as long a period that can be negotiated. The objective must be to keep stock, work in progress, debts to a minimum and creditors to a maximum.

Debtor days

This is calculated by dividing your trade debtors by annual sales and multiplying by 365. This shows the number of days' credit being afforded to your customers and should be compared both with your normal trade terms and the previous month's figures. Normal procedures should involve the preparation of a monthly-aged list of debtors showing the name of the customer, the value and to which month it relates.

The oldest and largest debtors can be seen at a glance for immediate consideration of what further recovery action is needed. The list may also show over-reliance on one or two large customers or the need to stop supplying a particularly bad payer until his arrears have been reduced to an acceptable level. Consideration should be given to making up bills to a date before the end of the month and making sure the accounts are sent out immediately, followed by a statement four weeks later.

Consider giving discounts for prompt payment. If all else fails, and legal action for recovery is being contemplated, call at the County Court and ask for their leaflets.

Stock turn

The level of stock should be kept to a minimum and the number of days' stock can be calculated by dividing the stock by the annual purchases and multiplying by 365. A worsening trend on a month-by-month basis shows the need for action. It is important to regularly make a full inventory of all stock and dispose of old or surplus items for cash. A stock control procedure to avoid stock losses and to keep stock to a minimum should be implemented.

Profitability

Whilst cash is vital in the short-term, profitability is vital in the medium-term. The two key percentage figures are the gross profit percentage and the net profit percentage. Gross profit is calculated by deducting the cost of materials and direct labour from the sales figures whilst net profit is arrived at after deducting all overheads. Possible reasons for changes in the gross profit percentage are:

- not taking full account of increases in materials and wages in the pricing of jobs

- too generous discount terms being offered
- poor management, over-manning, waste and pilferage of materials
- too much down-time on equipment which is in need of replacement.

If net profit is deteriorating after the deduction of an appropriate reward for your own efforts, including an amount for your own personal tax liability, you should review each item of overhead expenditure in detail asking the following questions:

- can savings be made in non-productive staff?
- is sub-contracting possible and would it be cheaper?
- have all possible energy-saving methods been fully explored?
- do the company's vehicles spend too much time in the yard and can they be shared or their number reduced?
- is the expenditure on advertising producing sales - review in association with 'marketing' above?

Over-trading

Many inexperienced businessmen imagine that profitability equals money in the bank and in some cases, particularly where the receipts are wholly in cash, this may be the case. But often, increased business means higher stock inventories, extra wages and overheads, increased capital expenditure on premises and plant, all of which require short-term finance.

Additionally, if the debtors show a marked increase as the turnover rises, the proprietor may find to his surprise that each expansion of trade reduces rather than increases his cash resources and he is continually having to rely on extensions to his existing credit.

The business, which had enough funds for start-up, finds it does not have sufficient cash to run at the higher level of operation and the bank manager may he getting anxious about the increasing overdraft. It is essential for those who run a business that operates on credit terms to be aware that profitability does not necessarily mean increased cash availability. Regular monthly management information on marketing and finance as described in this chapter will enable over-trading to be recognised and remedial action to be taken early.

If the situation is appreciated only when the bank and other creditors are pressing for money, radical solutions may be necessary, such as bringing in new finance, sale and leaseback of premises, a fundamental change in the terms of trade or even selling out to a buyer with more resources. Professional help from the firm's accountant will be needed in these circumstances.

Break-even point

The costs of a business may be divided into two types - variable and fixed. *Variable costs* are those which increase or decrease as the volume of work goes up or down and include such items as materials used, direct labour and power machine tools. *Fixed costs* are not related to turnover and are sometimes called fixed overheads. They include rent, rates, insurance, heat and light, office salaries and plant depreciation. These costs are still incurred even though few or no sales are being made.

Many small businessmen run their enterprises from home using family labour as back-up; they mainly sell their own labour and buy materials and hire plant only as required. By these means they reduce their fixed costs to a minimum and start making profits almost immediately. However, larger firms that have business premises, perhaps a small workshop, an office and vehicles, need to know how much they have to sell to cover their costs and become profitable.

In the case of a new business it is necessary to estimate this figure but where annual accounts are available a break-even chart based on them can be readily prepared. Suppose the real or estimated figures (expressed in £000s) are:

	%	£
Sales	100	400
Variable costs	66	265
Gross profit	34	135
Fixed costs	13	50
Net profit	21	85

Break-even point $= \dfrac{50 \text{ divided by (1 less variable costs \%)}}{\text{sales}}$

= 50 divided by (1 less 0.6625)
= 50 divided by 0.3375
= £148 (thousand)

In practice things are never quite as clear cut as the figures show, but nevertheless this is a very useful tool for assessing not only the break-even point but also the approximate amount of loss or profit arising at differing levels of turnover and also for considering pricing policy.

Taxation

The first decision usually required to be made from a taxation point of view is which trading entity to adopt. The options available are set out below.

Sole trader

A sole trader is a person who is in business on his own account. There is no statutory requirement to produce accounts nor is there a necessity to have them audited. A sole trader may, however, be required to register for PAYE and VAT purposes and maintain records so that Income Tax and VAT returns can be made. A sole trader is personally liable for all the liabilities of his business.

Partnership

A partnership is a collection of individuals in business on their own account and whose constitution is generally governed by the Partnership Act 1890. It is strongly recommended that a partnership agreement is also established to determine the commercial relationship between the individuals concerned.

The requirements in relation to accounting records and returns are similar to those of a sole trader and in general a partner's liability is unlimited.

Limited company

This is the most common business entity. Companies are incorporated under the Companies Act 1985 which requires that an annual audit is carried out for all companies with a turnover in excess of £1,000,000 or a review if the turnover is less than £1,000,000 and that accounts are filed with the Companies Registrar. Generally an individual shareholder's liability is limited to the amount of the share capital he is required to subscribe.

Advantages

In view of the problems and costs of incorporating an existing business, it is important to try and select the correct trading medium at the commencement of operations. It is not true to say that every business should start life as a company.

Many businesses are carried on in a safe and efficient manner by sole traders or partnerships. Whilst recognising the possible commercial advantages of a limited company, taxation advantages exist for sole traderships and partnerships, such as income tax deferral and National Insurance saving. No decision should be taken without first seeking professional advice.

The benefit of limited liability should not be ignored although this can largely be negated by banks seeking personal guarantees. In addition, it may be easier for the companies to raise finance because the bank can take security on the debts of the company that could be sold in the future, particularly if third-party finance has been obtained in the form of equity.

Self-assessment

From the tax year 1996/97 the burden of assessing tax shifted from the Inland Revenue to the individual tax payer. The main features of this system are as follows:

- the onus is on the taxpayer to provide information and to complete returns
- tax will be payable on different dates
- the taxpayer has a choice: he can calculate his tax liability at the same time as making his return and this will need to be done by 31st January following the end of the tax year. Alternatively, he can send in his tax return before 30 September and the Inland Revenue will calculate the tax to be paid on the following 31 January
- the important aspect to the system is that if the return is late, or the tax is paid late, there will be automatic penalties and/or surcharges imposed on the taxpayer.

Tax correspondence

Businessmen do not like letters from the Inland Revenue but they should resist the temptation to tear them up or put them behind the clock and forget about them. All Tax Calculations and Statements of Account should be checked for accuracy immediately and any queries should be put to your accountant or sent to the Tax District that issued the document.

Keep copies of all correspondence with the Inland Revenue. Letters can be mislaid or fail to be delivered and it is essential to have both proof of what was sent as well as a permanent record of all correspondence.

Dates tax due

Income Tax

Payments on account (based on one half of last year's liability) are due on 31 January and 31 July. If these are insufficient there is a balancing payment due on the following 31 January – the same day as the tax return needs to be filed. For example:

for the year 2000/01	Tax due £5,000 (1999/00 was £4,000) First payment on account of £2,000 is due on 31.01.01 Second payment on account of £2,000 is due on 31.07.01 Balancing payment of £1,000 is due on 31.01.02

Note that on 31.01.02, the first payment on account of £2,500 fell due for the tax year 2001/02.

Tax in business

Spouses in business

If spouses work in the business, perhaps answering the phone, making appointments, writing business letters, making up bills and keeping the books, they should be properly remunerated for it. Being a payment to a family member, the Inspector of Taxes will be understandably cautious in allowing remuneration in full as a business expense. The payment should be:

- actually paid to them, preferably weekly or monthly and in addition to any housekeeping monies
- recorded in the business book
- reasonable in amount in line with their duties and the time spent on them.

If the wages paid to them exceed £77.00 per week, Class 1 employer's and employee's NIC becomes due and if they exceed £4,155 p.a. (assuming they have no other income) PAYE tax will also be payable.

It should also be noted that once small businesses are well established and the spouses' earnings are approaching the above limits, consideration may be given to bringing them in as a partner. This has a number of effects:

- there is a reduced need to relate the spouse's income (which is now a share of the profits) to the work they do
- they will pay Class 2 and Class 4 NIC instead of the more costly Class I contributions and PAYE will no longer apply to their earnings but remember that, as partners, they have unlimited liability.

Premises

Many small businessmen cannot afford to rent or buy commercial premises and run their enterprises from home using part of it as an office where the books and vouchers, clients' records and trade manuals are kept and where estimates and plans are drawn up. In these circumstances, a portion of the outgoings on the property may be claimed as a business expenses. An accountant's advice should be sought to ensure that the capital gains tax exemption that applies on the sale of the main residence is not lost.

Fixed Profit Car Scheme

It may be advantageous to calculate your car expenses using a fixed rate per business mile. A condition is that your annual turnover is below the VAT threshold (currently £55,000). Ask your accountant about this. A proper record of business mileage must be kept.

Vehicles

Car expenses for sole traders and partners are usually split on a fractional mileage basis between business journeys, which are allowable, and private ones, which are not, and a record of each should he kept. If the business does work only on one or two sites for only one main contractor, the inspector may argue that the true base of operations is the work site not the residence and seek to disallow the cost of travel between home and work. It is tax-wise and sound business practice to have as many customers as possible and not work for just one client.

Business entertainment

No tax relief is due for expenditure on business entertainment and neither is the VAT recoverable on gifts to customers, whether they are from this country or overseas. However, the cost of small trade gifts not exceeding £10 per person per annum in value is still admissible provided that the gift advertises the business and does not consist of food, drink or tobacco.

Income tax (2003/04)

Personal allowances

The current personal allowance for a single person is £4,155. The personal allowance for people aged 65 to 74 and over 75 years are £6,610 and £6,720 respectively. The married couple's allowance was withdrawn on 5 April 2000, except for those over 65 on that date.

Taxation of husband and wife

A married woman is treated in much the same way as a single person with her own personal allowance and basic rate band. Husband and wife each make a separate return of their own income and the Inland Revenue deals with each one in complete privacy; letters about the husband's affairs will be addressed only to him and about the wife's only to her unless the parties indicate differently.

Rates of tax

Tax is deducted at source from most banks and building societies accounts at the rate of 20%. The rates of tax for 2000/01 are as follows:

Lower rate: 10% on taxable income up to £1,960
Basic rate: 22% on taxable income between £1,960 and £30,500
Higher rate: 40% on taxable income over £30,500

Dividends carry a 10% non-repayable tax credit. Higher rate taxpayers pay a further tax on dividends of 22.5%.

Mortgage interest relief

This is no longer available after 5 April 2000.

Business losses

These are allowed only against the income of the person who incurs the loss. For example, a loss in the husband's business cannot be set against the wife's income from employment.

Joint income

In the case of joint ownership by a husband and wife of assets that yield income, such as bank and building society accounts, shares and rented property, the Inland Revenue will treat the income as arising equally to both and each will pay tax on one half of the income. If, however, the asset is owned in unequal shares or one spouse only and the taxpayer can prove this, then the shares of income to be taxed can be adjusted accordingly if a joint declaration is made to the tax office setting out the facts.

Capital Gains Tax

Where an asset is disposed of, the first £7,900 of the gain is exempt from tax. In the case of husbands and wives, each has a £7,900 exemption so if the ownership of the assets is divided between them, it is possible to claim exemption on gains up to £15,800 jointly in the tax year. Any remaining gain is chargeable as though it were the top slice of the individual's income; therefore according to his or her circumstances it might be charged at 10%, 22% or 40%.

Self-employed NIC rates (from 6 April 2000)

Class 2 rate
Charged at £2.00 per week. If earnings are below £4,615 per annum averaged over the year, ask the DSS about 'small income exception'. Details are in leaflet CA02.

Class 4 rate
Business profits up to £4,615 per annum are charged at NIL. Annual profits between £4,615 and £30,940 are charged at 8% of the profit. There is a charge on profits over £30,940. Class 4 contributions are collected by the Inland Revenue along with the income tax due.

Capital allowances (depreciation) rates

Plant and machinery:	25% (40% first-year allowance is available for certain small businesses)
Business motor cars - cost up to £12,000:	25%
- cost over £12,000:	£3,000 (maximum)
Industrial build	4%
Commercial and industrial buildings in Enterprise Zones:	100%
Computers and software equipment	100%

THE CONSTRUCTION INDUSTRY TAX DEDUCTION SCHEME

General

The new Construction Industry Tax Deduction Scheme is known as the 'CIS' scheme and replaced the old '714' scheme. As the scheme operates whenever a contractor makes a payment to a sub-contractor, the businessman should visit his local income tax enquiry office and obtain copies of the Inland Revenue booklet IR 14/15 (CIS) and leaflet IR 40 which will explain the conditions under which the Inland Revenue will issue a registration card or (CIS6) certificate and precisely when the scheme applies.

Everyone who carries out work in the Construction Industry Scheme must hold a registration card (CIS4) or a tax certificate (CIS6). Certain larger companies use a special certificate (CIS5).

If the sub-contractor has a registration card but does not hold a valid tax certificate (CIS6) issued to him by the Inland Revenue, then the contractor *must* deduct 18% tax from the whole of any payment made to him (excluding the cost of any materials) and to account to the Inland Revenue for all amounts so withheld.

To enable the subcontractor to prove to the Inspector of Taxes that he has suffered this tax deduction, the contractor must complete the three-part tax payment voucher (CIS25) showing the amount withheld. These vouchers must be carefully filed for production to the Inspector after the end of the tax year along with the tax return. Any tax deducted in this way over and above the sub-contractor's agreed liability for the year will be repaid by the Inland Revenue. If he holds a (CIS6) certificate the payment may be made in full without deducting tax.

A small business that does work only for the general public and small commercial concerns is outside the scheme and does not need a certificate to trade. If, however, it engages other contractors to do jobs for it, the business would have to register under the scheme as a contractor and deduct tax from any payment made to a sub-contractor who did not produce a valid (CIS6) certificate. If in doubt, consult your accountant or the Inland Revenue direct.

VAT

The general rule about liability to register for VAT is given in the VAT office notes. It is possible to give here only a brief outline of how the tax works. The rules that apply to the construction industry are extremely complex and all traders must study *The VAT Guide* and other publications.

Registration for VAT is required if, at the end of any month, the value of taxable supplies in the last 12 months exceeds the annual threshold or if there are reasonable grounds for believing the value of the taxable supplies in the next 30 days will exceed the annual threshold.

Taxable supplies include any zero-rated items. The annual threshold is £55,000. The amount of tax to be paid is the difference between the VAT charged out to customers *(output tax)* and that suffered on payments made to suppliers for goods and services *(input tax)* incurred in making taxable supplies. Unlike income tax there is no distinction in VAT for capital items so that the tax charged on the purchase of, for example, machinery, trucks and office furniture, will normally be reclaimable as *input tax*.

VAT is payable in respect of three monthly periods known as 'tax periods'. You can apply to have the group of tax periods that fits in best with your financial year. The tax must be paid within one month of the end of each tax period. Traders who receive regular repayments of VAT can apply to have them monthly rather than quarterly. Not all types of goods and services are taxed at 17.5% (i.e. the standard rate). Some are exempt and others are zero-rated.

Zero-rated

This means that no VAT is chargeable on the goods or services, but a registered trader can reclaim any *input* tax suffered on his purchases. For instance, a builder pays VAT on the materials he buys to provide supplies of constructing but if he is constructing a new dwelling house, this is zero rated. The builder may reclaim this VAT or set it off against any VAT due on standard rated work.

Exempt

Supplies that are exempt are less favourably treated than those that are zero rated. Again no VAT is chargeable on the goods or services but the trader cannot reclaim any *input* tax suffered on his purchases.

Standard-rated

All work which is not specifically stated to be zero rated or exempt is standard-rated, i.e. VAT is chargeable at the current rate of 17.5% and the trader may deduct any *input* tax suffered when he is making his return to the Customs and Excise. If for any reason a trader makes a supply and fails to charge VAT when he should have done so (e.g. mistakenly assuming the supply to be zero rated), he will have to account for the VAT himself out of the proceeds. If there is any doubt about the VAT position, it is safer to assume the supply is standard rated, charge the appropriate amount of VAT on the invoice and argue about it later.

Time of supply

The *time* at which a supply of goods or services is treated as taking place is important and is called the 'tax point'. VAT must be accounted for to the

Customs and Excise at the end of the accounting period in which this 'tax point' occurs. For the supply of goods which are 'built on site', the 'basic tax point' is the date the goods are made available for the customer's use, whilst for *services* it is normally the date when all work except invoicing is completed.

However, if you issue a tax invoice or receive a payment before this 'basic tax point' then that date becomes a tax point. In the case of contracts providing for stage and retention payments, the tax point is either the date the tax invoice is issued or when payment is received, whichever is the earlier.

All the requirements apply to sub-contractors and main contractors and it should be noted that, when a contractor deducts income tax from a payment to a sub-contractor (because he has no valid CIS6) VAT is payable on the full gross amount *before* taking off the income tax.

Annual accounting

It is possible to account for VAT other than on a specified three month period. Annual accounting provides for nine equal installments to be paid by direct debit with annual return provided with the tenth payment. £300,000.

Cash accounting

If turnover is below a specified limit, currently £600,000, a taxpayer may account for VAT on the basis of cash paid and received. The main advantages are automatic bad debt relief and a deferral of VAT payment where extended credit is given.

Bad debts

Relief is available for debts over 6 months.

Part Four

GENERAL CONSTRUCTION DATA

GENERAL CONSTRUCTION DATA

The metric system

Linear

1 centimetre (cm)	=	10 millimetres (mm)
1 decimetre (dm)	=	10 centimetres (cm)
1 metre (m)	=	10 decimetres (dm)
1 kilometre (km)	=	1000 metres (m)

Area

100 sq millimetres	=	1 sq centimetre
100 sq centimetres	=	1 sq decimetre
100 sq decimetres	=	1 sq metre
1000 sq metres	=	1 hectare

Capacity

1 millilitre (ml)	=	1 cubic centimetre (cm3)
1 centilitre (cl)	=	10 millilitres (ml)
1 decilitre (dl)	=	10 centilitres (cl)
1 litre (l)	=	10 decilitres (dl)

Weight

1 centigram (cg)	=	10 milligrams (mg)
1 decigram (dg)	=	10 centigrams (mcg)
1 gram (g)	=	10 decigrams (dg)
1 decagram (dag)	=	10 grams (g)
1 hectogram (hg)	=	10 decagrams (dag)

Conversion equivalents (imperial/metric)

Length

1 inch	=	25.4 mm
1 foot	=	304.8 mm
1 yard	=	914.4 mm
1 yard	=	0.9144 m
1 mile	=	1609.34 m

Area

1 sq inch	=	645.16 sq mm
1 sq ft	=	0.092903 sq m
1 sq yard	=	0.8361 sq m
1 acre	=	4840 sq yards
1 acre	=	2.471 hectares

Liquid

1 lb water	=	0.454 litres
1 pint	=	0.568 litres
1 gallon	=	4.546 litres

Horse-power

1 hp	=	746 watts
1 hp	=	0.746 kW
1 hp	=	33,000 ft.lb/min

Weight

1 lb	=	0.4536 kg
1 cwt	=	50.8 kg
1 ton	=	1016.1 kg

Conversion equivalents (metric/imperial)

Length

1 mm	=	0.03937 inches
1 centimetre	=	0.3937 inches
1 metre	=	1.094 yards
1 metre	=	3.282 ft
1 kilometre	=	0.621373 miles

Area

1 sq millimetre	=	0.00155 sq in
1 sq metre	=	10.764 sq ft
1 sq metre	=	1.196 sq yards
1 acre	=	4046.86 sq m
1 hectare	=	0.404686 acres

Liquid

1 litre	=	2.202 lbs
1 litre	=	1.76 pints
1 litre	=	0.22 gallons

Horse-power

1 watt	=	0.00134 hp
1 kw	=	134 hp
1 hp	=	0759 kg m/s

Weight

1 kg	=	2.205 lbs
1 kg	=	0.01968 cwt
1 kg	=	0.000984 ton

Temperature equivalents

In order to convert Fahrenheit to Celsius deduct 32 and multiply by 5/9.
To convert Celsius to Fahrenheit multiply by 9/5 and add 32.

Fahrenheit	Celsius
230	110.0
220	104.4
210	98.9
200	93.3
190	87.8
180	82.2
170	76.7
160	71.1
150	65.6
140	60.0
130	54.4
120	48.9
110	43.3
100	37.8
90	32.2
80	26.7
70	21.1
60	15.6
50	10.0
40	4.4
30	-1.1
20	-6.7
10	-12.2
0	-17.8

Lighting

The formula for assessing the amount of illumination required is usually described as the lumen method and is expressed as:

$$F = \frac{A \times Eav}{CU \times M}$$

where required	F	is the total number of lumens
	A	is the area to be illuminated
	Eav	is the average illumination on the working plane
	CU	is the coefficient of illumination and
	M	is the maintenance factor.

The maintenance factor is usually stated as 0.8 but can be 0.6 in dirty areas. The figures to be applied to Eav are based upon tables of average illumination levels in different working conditions.

Area	**Lux**	**Lumens/ft2**
General office conditions	500	50
Drawing office	750	75
Corridors, store rooms	300	30
Shop counters	500	50
Watch repairing	3000	300
Proofreading	750	75
Living rooms	100	10
Bedrooms	50	5

To select the coefficient of utilisation (CU) it is necessary to determine the room index.

$$RI = \frac{L \times W}{Hm(L \times W)}$$

where	RI	=	is the room index
	L	=	is the length of the room
	W	=	is the width of the room and
	Hm	=	is the height of the fitting above the working plan.

A working height of 0.85m above floor level should be allowed where the working plane is a desk or bench top. The coefficient of utilisation (CU) is selected from the manufacturer's design information and after calculating the total illumination required, the design lumens can be chosen and the number of light fittings can be assessed.

$$\text{Number of light fitting} = \frac{\text{Illumination required (F)}}{\text{Design lumens per lamp}}$$

Index